Modelling, Monitoring and Diagnostic Techniques for Fluid Power Systems

John Watton

Modelling, Monitoring and Diagnostic Techniques for Fluid Power Systems

 Springer

John Watton, BSc, PhD, DSc, CEng, FIMechE
Institute of Machines and Structures
Cardiff School of Engineering
Cardiff University
Cardiff
CF24 3AA
Wales
UK

British Library Cataloguing in Publication Data
Watton, John
  Modelling, monitoring and diagnostic techniques for fluid
  power systems
  1.Fluid power technology 2.Machinery - Monitoring 3.Fluid
  power technology - Mathematical models
  I.Title
  620.1'06
ISBN-13:   9781846283734
ISBN-10:   1846283736

Library of Congress Control Number: 2006938341

ISBN 978-1-84628-373-4   e-ISBN 978-1-84628-374-1   Printed on acid-free paper

9 8 7 6 5 4 3 2 1

Springer Science+Business Media
springer.com

Preface

This book is aimed towards the condition monitoring and diagnostics aspects of fluid power control and is intended to replace and significantly extend my earlier book published in 1992, the majority of copies being destroyed in a warehouse flood shortly after its publication. This salutary lesson in condition monitoring did not persuade the publisher to re-print the book and I suspect that a connection between the topic and the event was never really made. The net result of this has been that a quite different book has emerged. This is perhaps not surprising given the large amount of work on the subject that I and others have done over the intervening years since my first book on the subject was conceived. In particular the application of condition monitoring and fault diagnosis to hydraulic systems has undergone a great deal of experimental validation in research laboratories and on real plant and I have taken much delight from my industrial collaborations. However, research and development on condition monitoring and fault diagnosis of fluid power systems is still not significantly popular in universities. This is not helped by the relative absence of fluid power as a mainstream subject at undergraduate level. This is quite remarkable when one thinks of the diversity of applications from primary materials processing and manufacturing to mobile machines, modern automotive and aerospace engineering. Quite simply there are a vast number of areas where fluid power is the only viable solution. I propose therefore that many undergraduate and graduate engineers are missing a most exciting subject area that involves:

- new materials technologies
- solid mechanics, fluid mechanics and thermodynamics
- component, systems and machine static and dynamic design
- modern control theory and computer control techniques
- artificial intelligence techniques
- new sensor technologies
- signal processing and algorithms
- component and systems modelling and simulation
- condition monitoring and fault diagnosis

It is rare that a project does not involve several of these aspects and a fluid power engineer really now needs a systems, perhaps mechatronics, approach to new challenges. It is probably fortuitous that since my interest in condition monitoring began in the late 1980s, there has been a fruition of many concepts being applied to real systems from my point of view, and this book is intended to bring these together.

Each chapter could be a book in its own right and this has presented a problem in trying to convey essential issues without becoming a research publication with all historical references. It is important to quote appropriate additional reading material but a line has had to be drawn regarding how much should be included; for example it is possible to include over 1000 references on transmission line theory alone. I have attempted to include important references with the view that other works tend, but not always, to be covered in the publications quoted. The reader will note some early references since formative work now tends to be ignored, or certainly not acknowledged, in modern publications.

Chapter 1 deals with the general background of the subject matter, illustrating the reason for condition monitoring and some general principles that apply or can be applied. Hopefully this sets out the need for the ensuing chapters which then discuss some of the important details, in my opinion.

Chapter 2 considers modelling and computer simulation with appropriate basic theory and its practical application. This chapter was particularly difficult to minimise to a sensible length but most of what is included is supported by practical results. It therefore forms a useful introduction to fluid power with some material taken from my first book on Fluid Power Systems published in 1989 but now out of print. The reader will note the absence of any detailed control theory, beyond some fundamental ideas, since this is not really necessary within the context of this book; also background theory in general is not exhaustive in this area for the same reason. More importantly, what has been included is material that I have found to be useful for real applications particularly from my work with the fluid power and manufacturing industries. The reader will note several applications of artificial neural networks and related ideas such as data-based modelling just hinted at in my previous book on condition monitoring.

Chapter 3 considers condition monitoring methods where the theory of Chapter 2 is put into practice. In addition, pragmatic concepts of signal monitoring and processing are included to illustrate the practical reality of combining a sensible amount of both theory and intuition, perhaps experience. Some new algorithms developed at Cardiff are introduced here but an important message is again the practical limitations of each method and the fact that in reality several approaches should be tried to give confidence in the emerging diagnostic. Some sensor information is covered although this is not an attempt to overview the general field of condition monitoring, but only what I have found useful for my contribution to the subject. Perhaps the main themes of this chapter are pressure and flow monitoring, dynamic data analysis including vibration, and oil/wear debris analysis. This chapter also considers expert systems and knowledge-based reasoning. This is a fascinating area, quite complex and useful for situations such as multiple fault conditions, but is still enjoying only a modest evolution. It is intended to show how rules may be developed from some rather basic theoretical concepts, which then actually give a great deal of information on the probable fault state of the hydraulic circuit. Again many practical examples are used to illustrate the concepts, from simple drives and lifting systems to a seven-stand steel strip finishing mill.

Chapter 4 gives many examples of component faults in pumps and motors taken from industrial sources on the cause and solutions for breakdowns that may occur in a hydraulic circuit. It considers many components and typical failures that have been deduced over many year of experience.

A comprehensive list of books, papers and further reading is included together with a detailed Index.

Finally I must thank my many industrial contacts and friends who have freely given information and provided funding and equipment that has helped me progress this subject matter. Many of my excellent PhD and EngD research students have contributed to the contents of this book and I congratulate them not only on their work but also on their common sense in selecting such an important research topic.

Cardiff 2005   *J. Watton*

Contents

Chapter 1

Introduction

1.1 Why Implement Condition Monitoring?

The rapidly changing industrial climate now almost demands that each facet of operation be closely examined with a view to, broadly, obtaining:

- the highest product quality
- efficiency optimisation
- improvement in the safety of operation
- maximum profitability

By way of an example consider a steel billet forging press system shown in Figure 1.1 with the hydraulic circuit schematic shown in Figure 1.2. These presses vary in complexity, particularly multi-axis systems, and the digital control approaches incorporate advanced monitoring and diagnostic support.

Considering one issue of many in such complex circuits, a pump failure will inevitably produce debris that could be carried to the press cylinders resulting in financially-damaging consequences:

- The cost of pump removal, pump repair/replacement, pump re-fitting is often tolerable.
- Main press cylinders can be up to 2 m bore diameter and the cost of replacement or refurbishment will be significantly higher than the pump and probably prohibitive in the case of a new cylinder; the cylinder may well be manufactured in another country with additional shipping implications. Whatever the course of action taken, a significant loss of press operation time will occur.
- Downtime losses in the case of a minor pump problem, due to material re-heating and lost production will still be financially damaging to the press operator.
- The loss of component supply to meet the demands of the end user could have consequential financial penalties to the press operator, and to a point that could be financially crippling.

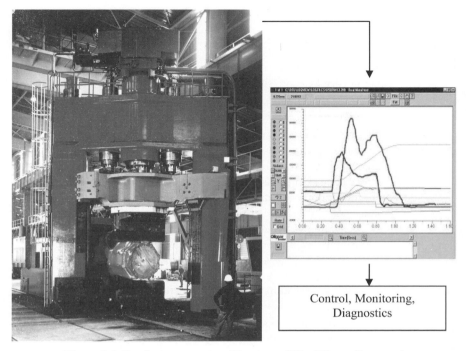

Figure 1.1. Forging press system (Courtesy of The Oilgear Company)

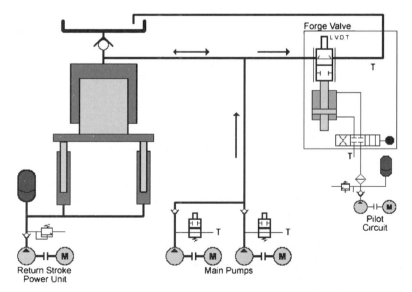

Figure 1.2. Forging press hydraulic circuit schematic (Courtesy of The Oilgear Company)

Clearly with systems such as this the main pumps must have specially designed debris filtering units to protect the circuit. Filtration is a key issue in fluid power and leads to perhaps the first rule for a hydraulic system:

> At the very least – provide adequate filtration and protect components from debris damage

Manufacturing profitability and efficiency are clearly connected, and in this context condition monitoring can play a vital part particularly when linked to a condition-based maintenance (CBM) policy. By monitoring the plant "health", potentially disastrous failures can be avoided, a corollary being that safety is also improved.

In addition to this, more information can be obtained relating to the plant operation, a by-product not always fully appreciated when embarking on a new monitoring venture. New, often unexpected, faults occur and information can often be acquired that contributes to improving the machine/plant operation. Unacceptable working practices can also be detected such as operating machines at incorrect speeds, sometimes near critical shaft frequencies. This is particularly aided by on-line computer-based monitoring where operators and technicians can readily see unacceptable trends.

The consequences of component failure, operator error, or lack of system integrity is sadly too well known in the public domain. Although the number of incidents are low when compared with the activity, the resulting loss of life is perceived as unacceptable when it occurs in areas such as air transport, chemical processing, and the energy supply industry. Condition monitoring and protection for safety-critical applications is now a necessity.

However, condition monitoring by itself will never eliminate major failures unless it is embraced within a "total quality" approach, particularly in industrial applications. This expression of total quality has evolved via the Japanese manufacturing industry and considers customer requirements, continuing education and training of the workforce and management, in addition to the technological aspects of the manufacturing operation. This total quality approach also appears under different descriptions such as Plant Asset Management and Total Product Maintenance, the latter representing yet another Japanese initiative in this field. Responsibilities are devolved to individuals as well as groups associated with areas of production.

Vast amounts of money, in some cases up to 15% of companies' sales, are spent maintaining assets. This clearly cannot be allowed to continue and Total Product Maintenance is proposed as a necessary approach to remain competitive in manufacturing industries.

There are various aspects that need to be considered ranging from the simplest of operator tasks through to advanced monitoring techniques and ease of maintenance when considering both existing and new designs. The simplest of tasks could be inspection for oil leaks, audible changes in noise levels or vibration, awareness of working temperatures etc., and these may be easily logged by the operator. Total Product Maintenance therefore represents an overall philosophy whereby all aspects are considered, success requiring a positive attitude from all concerned in the industry and with a high degree of motivation.

Another way of looking at a manufacturing operation is to study other companies with a view to defining the "Best World Practice". This may actually result in companies from different parts of the world agreeing to co-operate on such a scheme with reciprocal exchange arrangements being established. Once the Best World Practice has been defined, the company then examines its own procedures and then takes steps to improve where necessary. The paradox established by participating companies is of course resolved if they are supplying to different world markets and thus not in direct competition – this is now rapidly changing.

Modern industrial systems incorporating various subsystems tend to be highly interactive and hence a single component failure may have serious financial consequences. Condition monitoring is slowly replacing the common practice of regular preventative maintenance whereby components are replaced at pre-determined intervals before failure occurs – although it may not be impending. Even if the components are replaced there is a significant downtime with resultant costs and loss of revenue.

Failures inevitably occur at the most inconvenient time creating both technical, organisational and financial restrictions that could be minimised with condition monitoring. However, this suggests continual monitoring and a compromise has to be reached regarding whether it should be carried out hourly, daily, weekly or monthly. There is inevitably an element of experience required here since a manufacturing system could involve the monitoring of a large number of parameters resulting in a vast amount of data that has to be carefully analysed. This suggests the use of computer-based techniques, although there are now many powerful hand-held items of instrumentation that may well be the preferred option.

Fluid power systems often form a part of the total industrial operation and it is perhaps the area which is currently receiving the least attention, from a monitoring point of view. Fluid losses alone in such areas as mining and steel processing can result in hundreds of thousands of euros in replacement costs, apart from costs due to resulting failures or inefficiency of operation. It will therefore probably be consequential that fluid power systems will be monitored from the desire to initially concentrate on other components.

It is unfortunate that company boardroom decisions are often made on the basis of financial "payback" only, and this may be as short (or as sensible?) as 2 years. However, it is common experience, certainly that of the writer, that condition monitoring with its inherent "spin-offs" always exceeds the investment payback expectation.

When considering the cost of introducing CBM it is instructive to consider total maintenance costs as plant availability is improved due to reduced downtime resulting from the introduction of CBM. Figure 1.3 illustrates this point, in principle rather than detail, by the addition of reduced downtime costs with increased maintenance costs.

It follows from Figure 1.3 that increasing plant availability may well lead to increased costs at the higher end of the availability spectrum.

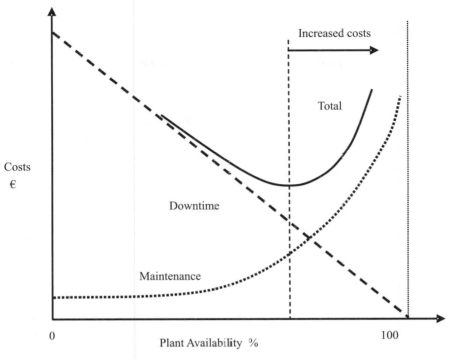

Figure 1.3. Maintenance/downtime cost variation

However, such a graph can only be derived by experience over many years of plant operation. The minimum cost point is also sensitive to individual cost fluctuations due to, for example, the current economic situation, world market share etc. Nevertheless, it seems to be common practice that companies continually strive to increase plant availability.

1.2 Three Maintenance Strategies

a) Breakdown Maintenance is the simplest of strategies to adopt since the plant is allowed to run without any rigorous supervision until it fails. Appropriate components are then repaired or changed on this ad hoc basis. This approach may actually be satisfactory for some subsystems such as a small pump circuit where often a standby pump is switched in, allowing the faulty pump to be repaired. Unfortunately there are more disadvantages than advantages of a breakdown maintenance approach when industry in general is considered. Some issues are as follows:

- Breakdowns often occur at the most inconvenient time creating undesirable disruption to operation.
- Mobile machine applications, for example, may result in considerable delay until replacement components are found, transported to site, and installed. This

is costly to the operating contractor and also introduces potential litigation costs, for example in large civil engineering operations such as road construction.

- The failure of a single component, for example a pump, can result in undesirable metallic particles being transported to other parts of the circuit and may cause further problems or even additional failures. This will almost certainly create high replacement costs.
- In a manufacturing system there will be critical components that need to be replaced quickly. This suggests that replacements must be held in store and represents an undesirable addition to capital expenditure.
- The replacement of a faulty component may not be carried out correctly. This can result in a further failure occurring rather more quickly than expected and is not helped by the absence of monitoring. This may be particularly the case for bearing and/or gearbox component replacements.
- Unexpected breakdowns may be a safety hazard in critical areas of operation and it may be necessary to install additional safety protection equipment or components which introduces additional costs.
- The maintenance effect/staffing will be irregular, and it is difficult to arrange a policy that utilises the manpower available and in an efficient manner. There will be a period of relative inactivity followed by a period of intense activity perhaps stretching the manpower resources available.

b) Preventative Maintenance represents a distinct improvement on breakdown maintenance. A strict maintenance schedule is established whereby components are replaced at pre-determined intervals, these intervals being established using a combination of manufacturers' data and operational experience. There can still be high costs associated with such an approach since the failure characteristic of a component can only be defined statistically. Figure 1.4 shows a typical failure distribution characteristic and illustrates three distinct regimes of totally safe operation, probable breakdown, and guaranteed breakdown.

Some important issues are as follows:

- Maintenance is planned in advanced, no matter what the plant condition is, and often the plant is shut down for perhaps one/two days in the case of a steel mill, such that the maintenance team (often sub-contracted) can be deployed.
- Clearly there are costs associated with premature replacement where breakdown will probably not occur. Components removed may still be functional and are often refurbished.
- There are also costs associated with a delayed replacement which, statistically, results in a higher system availability but with a potentially high breakdown cost.
- In situations where, in principle, failure is simply not allowed to occur, such as in aerospace applications, then the additional costs of premature replacement must be tolerated. In some applications this cost may be greater than that incurred due to a failure: experience must decide which approach is preferable.

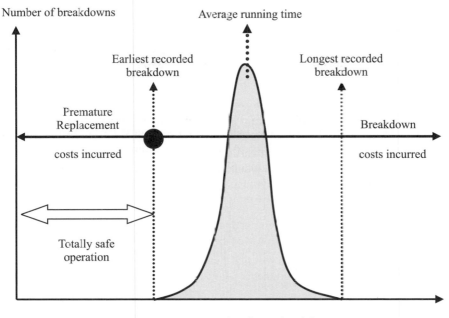

Figure 1.4. Typical component breakdown characteristic

- Experience, however, has also shown that the process of component replacement, often with a refurbished component that has previously been used, may lead to an unexpected premature failure. The reasons for this are varied, such as incorrect repair, oil contamination, incorrect re-assembly etc.
- However, there is no doubt that planned preventative maintenance is generally preferable to breakdown maintenance.

c) Condition-based Maintenance, incorporating condition monitoring with fault diagnosis in many cases, is generally now considered necessary for the optimum operation of modern plant.

- Condition-based maintenance involves the acquisition and analysis of data followed by some form of signal processing, further analysis and a decision-making policy. It may initially develop from the need to monitor just a few components, but inevitably it is expanded to cover complete systems.
- It is quite common for data to be obtained using hand-held instruments that are connected to appropriate test points around the plant. These may well be advanced electronic processing units where microcomputer technology has drastically reduced purchase cost. Data may then be transferred to a computer for analysis and trending of the appropriate parameter with operating time.
- On-line data acquisition provides the greatest flexibility since computer graphics combined with audible alarms make faults rapidly known to the operator.

Multi-channel systems are now commonplace and make the transition to on-line monitoring relatively easy to accommodate.

- The cost has to be carefully considered since computer hardware/software costs must be taken into account together with personnel training, system calibration and maintenance costs.

Consider, for example, the monitoring of a high-pressure axial piston pump. One parameter of interest is the case drain leakage flow rate, a natural feature of such a pump. A flow meter could easily be connected to the drain line when required and the flow rate could be checked to ensure that excessive leakage is not occurring. Figure 1.5 shows this set up with a possible measurement trend with time.

Some practical issues that need to be considered are:

- Experience will decide the flow loss value that indicates an unacceptable condition, the pump will then be removed and examined. A knowledge base is automatically developed and this allows fault indicator thresholds to be changed.
- A measurement such as this will not identify the cause of the increased leakage and more advanced measurements using pressure and/or vibration transducers

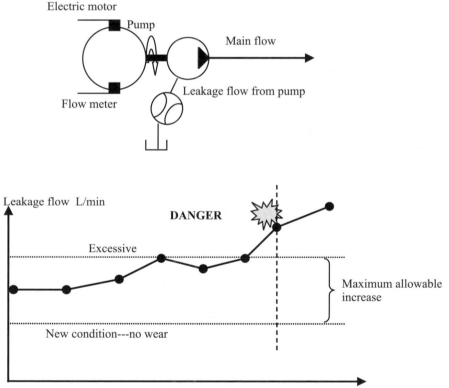

Figure 1.5. Leakage flow rate monitoring of an axial piston motor

together with oil contamination analysis may be necessary to pinpoint the actual fault.

- Off-line methods such as this are based upon cost effectiveness and the likelihood of the component failing together with additional consequences such as system failure and lead-time to repair.

To be able to make diagnostic predictions more measurements will inevitably be required thus increasing instrumentation costs and data analysis requirements. This suggests on-line monitoring with transducers hard-wired to a microcomputer which will have software written around the system in question. Pressure, flows, temperature etc., could be monitored and simply displayed on the computer screen with audible alarms being triggered as conditions deteriorate to a pre-determined unacceptable level.

More advanced on-line diagnostic applications are now implementing expert system concepts. Data is acquired in exactly the same way as previously described, and this data is then analysed using knowledge bases established within the expert system software. Various facts may then be interpreted using a set of rules that may be expanded as knowledge is gained from theoretical and operational experience. These are not explicit mathematical rules in the traditional sense as will be discussed later. An additional feature of this approach is that transducer integrity may also be checked in some cases since it cannot always be assumed that each transducer is indicating the correct parameter value.

Hence by incorporating a range of knowledge bases and analysis tools, a better estimate can be made of the system condition and the possible cause of a change in its condition. This is illustrated schematically in Figure 1.6.

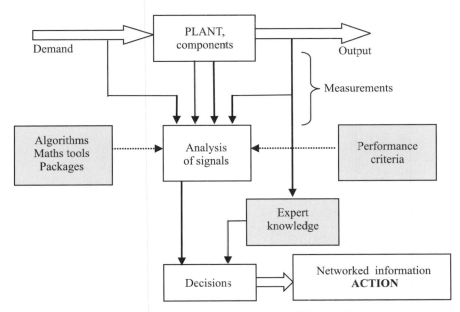

Figure 1.6. Expert system concepts for condition monitoring

1.3 Some Preliminary Conclusions

Some clearly defined steps may now be established when condition monitoring is being considered:

- The monitoring of complex fluid power systems will require some knowledge of the way components behave in that system. This may require characteristics and the way they change as components wear or, for example, as leakages occur between them.
- Understanding the performance of a system implies good design and the ability to predict the performance using appropriate computer software or CAD techniques.
- It is also important at the design stage to build appropriate monitoring points into the system so that sensors may be fitted at the building stage or later during operation. Sensors may therefore form a permanent feature or may simply be fitted when required if portable monitoring equipment is used.
- A monitoring strategy must be established together with presentation of data best suited to the application.
- It should be stated that whatever approach is adopted, there is generally no quick-and-easy solution to condition monitoring and fault diagnosis when applied to a large manufacturing plant.
- A range of techniques is available, some embracing advanced information technology. Commitment to detail is still required for technical, operational and management aspects.
- Improvements should be continually sought and it should also be realised that the dramatic improvements made by industry in recent years have not been instantly achieved. Investment payback can be as little as a few months, as experienced by the author, but may well be two to three years before the benefits of a new investment are fully realised.
- It has been suggested that an investment of typically 1% of the capital value of the plant is required, perhaps rising to 5% where safety risks must be eliminated.

1.4 Potential Benefits of CBM

There are many potential benefits of condition-based maintenance that have emerged over recent years, and may be broadly classified as follows, and not in any order of significance:

i) Reduced repair time and costs

- In some cases up to 80% reduction
- Labour and parts costs reduced with advanced knowledge
- Planned repairs less prone to problems than hurriedly-done repairs

ii) Improved plant operational knowledge

- Monitored data tends to reveal new fault information
- Monitored data can reveal incorrect plant operation due to either deliberate operator mis-use or the plant running at near-critical conditions

iii) A maintenance cost saving

- In some cases spares inventories have been reduced by up to 30% since advance warning allows just-in-time purchasing
- Priority action only on machines that need repair
- Actual number of failures reduced
- Planned maintenance reduced or possibly eliminated
- Reduced downtime hence reduced costs
- Maintenance teams arranged in advance and to suit the work needed

iv) Minimised revenue loss

- Impending failures detected and tracked, repairs carried out at a convenient and planned time
- Plant availability maintained or improved
- Revenue may actually increase with evidence of as much as 30%

v) Maintained product quality

- Monitored process parameters allow intelligent strategies to maintain product quality
- Causal effects on product quality changes can be tracked

vi) Improved plant life

- Serious damage avoided or minimised
- 'Knock-on' effects of damage transferred to other components avoided
- Longer running times for components normally changed under a planned maintenance scheme gives extended plant life

vii) Improved safety assurance, reduced personnel risk

- Data available to show improved performance, particularly on safety-critical aspects
- A demonstrable safer working environment gives added confidence to personnel
- Hospital and litigation costs reduced
- Improved safety assurance leads to minimised insurance premiums

viii) Improved plant design and operation

- CBM often produces much more information on plant behaviour and which parameters are crucial to efficient plant operation. This can often be valuable to the system designer and may lead to improved designs
- The monitored performance may also lead to improved ways of actually running the plant. In addition, the new databases developed can indicate whether or not unacceptable operating conditions are being approached

ix) Maintained customer relationship

- Maintenance of supply ensures customer satisfaction
- Maintenance of product quality aids customer satisfaction

1.5 Benefits Applied to Plant Economics

Output revenues, estimated over the plant life, must be compared against initial capital investment costs and daily operating costs as shown conceptually in Figure 1.7. Inflation, market price changes, revenue changes, planned scheduled maintenance, over the operating plant life has been neglected.

The effect of plant downtime is shown in Figure 1.8 and it can be seen that there are two major aspects, increased total costs and lost production revenue. Note that the lost profit interval is much greater than the downtime interval.

Once production has been lost, the investment return for the process can only be regained by either improving efficiency and/or extending plant operating life.

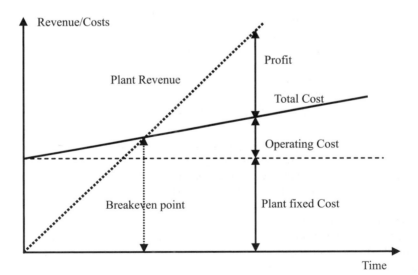

Figure 1.7. Ideal plant economics

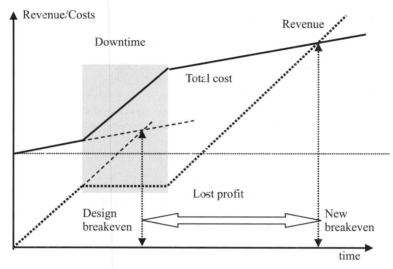

Figure 1.8. Effects of downtime on plant economics

A condition monitoring approach will then probably be crucial in meeting these new objectives.

There are many other issues to be faced at the same time such as switching to other production systems, the use of outside contract work, logistics of maintenance, all of which must be judged against a possible further downtime occurrence at a later date.

1.6 Types of Condition Monitoring Systems

The selection of the correct data acquisition approach is important and a number of possibilities exist depending on the complexity of the plant and the projected investment cost perceived as appropriate to the new condition monitoring strategy.

i) On-line multi-point systems
These are the most advanced data acquisition and signal processing systems and are usually 'hard wired' to the plant. These systems have the highest investment cost but offer a great deal of flexibility both for extending the number of monitoring points and adding new software fault diagnostic techniques. Computer terminals can be placed in appropriate plant offices giving direct access to a number of personnel. Fast transients may be captured allowing modern signal processing and identification to be used. Multi-point measurements are usually multiplexed to minimise cost.

ii) Surveillance systems
Data are again acquired on-line but the signal processing time is not so significant as with high-speed systems. They tend to be single-component dedicated systems (for example a tank level measurement), with high integrity, relatively low cost, and probably also hard-wired to the central monitoring system.

iii) Manual data collection systems

Data are acquired by handheld units from either single transducer monitoring points or multi-channel collection points around the plant. These handheld units can contain as much advanced processing power as many on-line computer systems and are capable of advanced diagnostics as well as simply recording quasi-static information such as temperatures, mean noise levels, etc. They can also be programmed to indicate the site route and which monitoring point is next on the route.

1.7 Methods of Condition Monitoring

Sensors are continually being developed, but the approaches may be broadly classified within the following areas:

i) Human – the senses – hearing, smell, touch (for example, temperature, vibration), are still quite common but based upon experience of working alongside the equipment or plant. Most of these "senses" may be replaced by "sensors".

ii) Steady-state measurements – changes in speed, torque, pressure, flow rates, temperature, vibration/noise dB readings, shaft alignment via laser measurement, etc. Process "outputs" such as product quality, throughput etc.

iii) Dynamic signal processing

- **Frequency analysis**
  A signal is analysed in terms of its frequency content allowing specific characteristics to be identified, usually from expected values such as rotation frequency of a machine. If the level increases at that known frequency then a changing condition is assumed. Vibration, in particular, may be measured with either accelerometers or acoustic emission sensors. In all cases, signal processing algorithms are required to convert the time signal into its frequency components. The usual method is to use the Fast Fourier Transform (FFT) algorithm. This is now standard software used in vibration measuring systems – simply measure and plot the results.

- **Time domain analysis**
  Methods are now available to actually characterise the time-varying signals. For example, step response tests or frequency response tests are well known from control theory, and apply to situations where known disturbances are applied and the output measured. However, new methods can work on the small fluctuations naturally occurring, and an important technique here is Time Encoded Signal Processing (TESP) analysis. Recent developments now embrace Artificial Neural Networks (ANNs) for data classification.

iv) Wear debris/fluid contamination analysis – particles are generated under wear or fault conditions, and any increase in particle generation rate down to micron size levels can now be detected either on-line or by sampling off-line. Also, the particle type can be determined to indicate from which part of the machinery it probably originated.

v) Fluid leakage detection – this may be simply visual. More advanced approaches will use on-line flow sensors or other sensors, such as pressure, which may serve to indicate flow losses via further systems analysis.

vi) Thermography – this is quite an expensive "camera" technique that converts the radiation spectrum into a colour coded pattern using infrared detection. It is a valuable tool for hot-spot detection, particularly in large-scale production systems utilising high-energy machines and processes.

vii) Corrosion – this occurs, for example, due to either:

* environmental effects
* internally within fluid components (such as pumps/pipelines, control valves etc.) due to chemical effects
* due to combustion chemical reactions
* bacteria effects, particularly in water-based fluids

viii) Erosion of a material can also occur due to localised fluid cavitation effects:

* Cavitation must be minimised by careful component and systems design, and it may be necessary to monitor a component if cavitation could be a problem.
* Sensors, particularly acoustic/stresswave, may be used for detecting incipient cavitation.

1.8 Failure Modes and Effects Analysis (FMEA)

When considering a large plant operation, with its inevitable distribution of faults, a methodological approach is required that seeks to prioritise the order in which faults are investigated and resolved. The FMEA approach does this by combining carefully considered fault data and the experience of the plant operations team. It is a process requiring continual investigation and action, and often leads to improved knowledge about the behaviour of the plant. The FMEA process in practice is typically as shown in Figure 1.9.

The FMEA method is a structured approach to fault diagnosis, fault correction, quality improvement, and has the following main advantages:

* it aims to recognise and evaluate the *actual* and *potential* failure modes
* it aims to recognise the cause of the failure modes
* it identifies sections that could eliminate or reduce the chance of failure
* it documents the corrective process

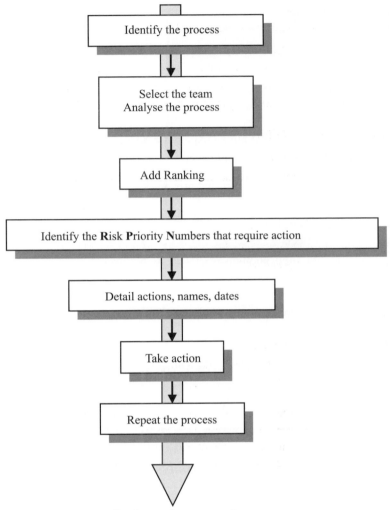

Figure 1.9. Essential steps in the FMEA approach to fault prioritisation

Now consider some details of the FMEA process:

i) Potential failure modes
It is important to continually ask the question – "what can go wrong"? It is also important to recall two key aspects:

- A potential failure mode is the way in which a process *could fail* to achieve objectives defined in the process description.
- What are the ways it might fail *not just the ways it has failed*.

ii) The effects of failure

What happens when the failure mode occurs?

- Effects could be defined as the effects on the customer, i.e. what they may notice or experience as a result of the failure mode when considering "output".
- Alternatively for system design, effects may be just the implication on the manufacturing process itself – perhaps the first steps for CBM.

iii) Potential causes of failure

- Define how the failure mode could occur.
- List every conceivable failure cause possible for each failure mode.

iv) Current Controls

These are the controls that either:

- prevent the failure mode occurring
- detect the failure if it occurs

v) Delta

This symbol is used if a potential failure mode has **Safety Implications**.

vi) Risk Priority Number RPN

This is the crucial FMEA indicator that ranks the severity of failure for the process. It is calculated as follows:

$$RPN = Severity \times Occurrence \times Detection \ Rank$$

The highest RPN is most significant and indicates where action is a priority. The RPN is continually evaluated as the process operation becomes more refined and of course improved.

vii) Severity

1	Of minor nature, not detectable.
2, 3	Will probably be noticed.
4, 5, 6	Causes some dissatisfaction, degradation of further processes.
7, 8	High degree of dissatisfaction and affect on further processes.
9, 10	Very high degree of dissatisfaction and severely affects further processes, safety critical areas.

viii) Occurrence

1	Remote, failure unlikely.
2	Very low.
3	Low.
4, 5, 6	Moderate, experience shows that the process occasionally fails.

7, 8   High, experience shows that the process often fails.
9, 10   Very high, failure almost inevitable.

iv) Detection
1, 2   Very high, failure almost certainly detected.
3, 4   High, good chance of detection.
5, 6   Moderate, may detect failure.
7, 8   Low, poor chance of detecting failure.
9, 10   Very low, no detection.

Consider an example taken from an actual manufacturing plant that represents just a 27-week FMEA study during which 18.3 hours of delays were recorded for the process operating 24 hours per day and 7 days per week. Data for this study are as follows:

		Downtime (min)	Downtime (%)	Occurrence
1	Hydraulic leaks	780	70.9	7
2	Loss of control	130	11.8	2
3	Servovalve failure, unstable, vibrating	50	4.5	2
4	Failure of actuators	15	1.3	1
5	Failure of pressure transducers	40	3.6	1
6	Cooling water pump failure	25	2.3	1

Problems associated with hydraulic leaks would appear to be the obvious priority area with perhaps loss of control following. However, a more detailed FMEA approach is needed to consider occurrence, severity, and detection rank as previously outlined.

Analysis of the faults leads to the following points:

- Faults **1, 2, 3** contribute **87.2%** of the downtime recorded.
- In the case of unstable control systems, the working life of the actuators can be reduced as seals wear at a higher rate. In addition hydraulic oscillations are undesirable when product quality is considered.
- The pressure transducer's life span is reduced when continuous pressure oscillations occur, and the zero pressure calibration often suffers.
- Oscillations lead to increased wear of servovalve components. This also occurs if the oil is contaminated with particles or with water if seals fail and allow process water into the hydraulics.

Severity, Occurrence, Detection, and the **RPN** are assessed by the FMEA team as follows:

		Severity	Occurrence	Detection	RPN
1	Hydraulic leaks	8	7	8	**448**
2	Loss of control	10	2	1	20
3	Servovalve failure, unstable, vibrating	10	2	8	**160**
4	Failure of actuators	8	1	9	72
5	Failure of pressure transducers	10	1	9	90
6	Cooling water pump failure	10	1	1	10

> **Action** on **leakages** as the **first priority**
> **Action** on **servovalves** as the **second priority**

These key actions should be implemented by the appropriate maintenance team and the plant performance re-evaluated at the agreed operating interval. Of equal importance is the assessment of the monitoring methods and whether improvements or new installations are necessary to minimise, hopefully eliminate, future plant downtime.

1.9 Correcting the Fault – Fault Tree Analysis

Having determined the most significant fault and taken action to replace the component or subsystem, an equally important requirement is to determine the cause of the fault. This can be a complex interactive problem in fluid power systems and a systematic method of cause and effect is needed to aid the action.

Fault Tree Analysis can help here in the sense that it provides a logical approach providing that all possibilities are included.

One approach, for example, could be to consider the observed characteristic of the deteriorating system and then work through the fault tree to determine the most probable cause of the fault. The concept of basic fault tree construction is shown in Figure 1.10, which illustrates just the beginning of a particular application.

The fault tree shown is by no means complete or definitive, but does suggest that expert input is needed to ensure that the major problems and means of correction are covered. Equipment manufacturers often provide manuals to aid this procedure prior to return for specialist analysis or repair. The systematic fault detection procedure can often be particularly advantageous in identifying not only component deficiencies but operating malpractice.

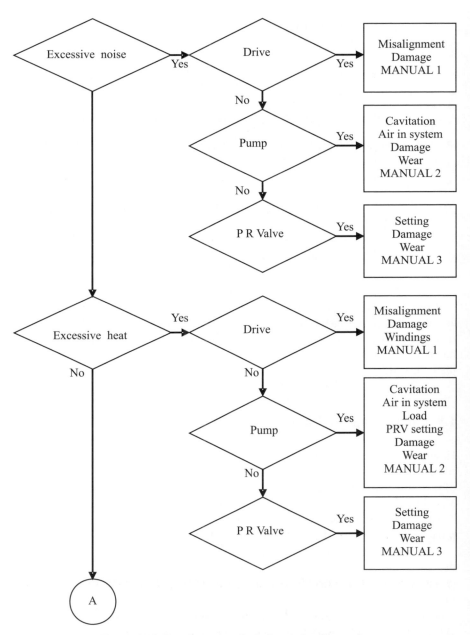

Figure 1.10. Development of a fault tree for diagnostics

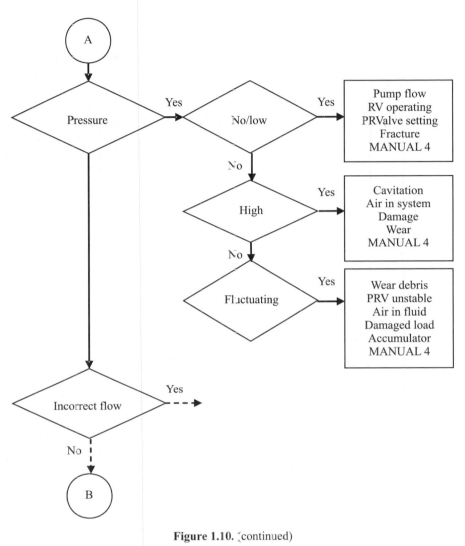

Figure 1.10. (continued)

1.10 Computer Simulation as a Fault Synthesis/Detecting Tool

The concept of a computer model-based approach to fault systhensis and/or detection is attractive but of course it does require sufficiently accurate component models. The understanding of both the steady-state and dynamic behaviour of fluid power components dramatically increased during the second half of the 20th century to date. However, one issue is simply the vast number of components on the market. A particular component, for example a pressure relief valve, has many variations of operating concept, size and slight design changes between each manufacturer. There is value in the approach for a particular industrial system that uses the same standard component replacements and where a good system model has been developed from extensive theoretical and experimental testing.

What does make the approach worth pursuing is also the ability to simulate faults and observe changes in system behaviour in a manner that simply could not be achieved in practice, for example pipe leaks. It is also particularly useful if system parameter changes, resulting from a fault, are measurable easily in practice. Some important features of a computer model-based approach are as follows:

- A fault may result in a change in steady-state pressures, flow rates, torque, speed, force at other points in the circuit.
- The steady-state energy distribution might be different at other points in the circuit and implies measurement of (force.distance), (pressure.flow rate), (torque.speed), (voltage.current).
- The dynamic performance may change, which implies the use of fast-acting transducers in practice to track such changes for a particular actuation sequence that is often used.
- A post-operation matching technique may be used whereby parameters in the computer simulation are changed in some statistical sequence such that the computer model behaviour emulates the practical system behaviour in the presence of a fault.

Perhaps the most obvious approach is to consider the steady-state behaviour of a system in its normal operating mode and with various faults. Changes in performance parameters may be deduced either by experience or from a well-developed and sufficiently accurate simulation model. For example consider the valve-controlled hydrostatic transmission shown in Figure 1.11 which was studied by Bull et al. (1997). In this study bond graph concepts were used to determine the particular parameter that best defines the energy fluctuation within the bond graph. An algorithm was developed to deal with qualitative behaviour and the associated energy storage and resistive elements representing losses due to leakage, friction, etc were included in the bond graph for steady state conditions. The outcome is a failure mode identifier such as "supply pressure up", "supply pressure down" etc. The first stage of this analysis method is to identify the potential failure modes and this can be done using the bond graph dynamic model. By considering faults, the power fluctuation is propagated around the circuit and the rules developed result in, for example, the concern for the motor which identifies a drop in speed.

In this way a failure mode table can be generated such as that shown in Table 1.1. Such a table depends upon the choice of a sufficiently significants failure level injection and it may well evolve that different faults create the same failure modes as evident from similar parameter changes. It was noted that fault simulation times were typically of the order of 1 second. The result of such an analysis may then require a more detailed investigation, for example via a complete dynamic simulation.

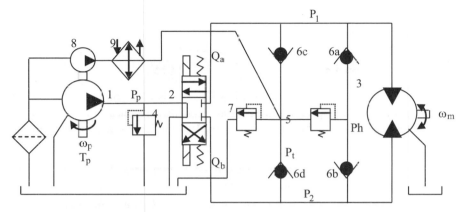

Figure 1.11. Valve-controlled hydrostatic drive [Bull, Stecki, Edge and Burrows, 1997]

Table 1.1. Failure modes and effects on the hydrostatic transmission

Failure Mode	Effect					
	T_p	ω_p	P_p	Q_a	Q_b	ω_m
Pump shaft slows down	up	down	down	down	down	down
Pump pressure falls	down	up	down	down	down	down
Pump pressure increases	up	down	up	up	up	up
Directional valve partially open	up	down	up	down	down	down
Increased friction in motor	down	up	down	up	down	down
Reduced line pressure	down	up	down	up	down	down
Reduced flow in RV 4	down	up	down	up	down	down
Increased flow in RV 4	up	down	up	down	up	up
Higher pressure in RV5	up	down	up	down	up	up
Lower pressure in RV5	down	up	down	up	down	down
Higher pressure in RV7	up	down	up	down	up	up
Lower pressure in RV7	down	up	down	up	down	down
Reduced flow in check valve 6a	down	up	down	up	down	down
Reduced flow in check valve 6b	down	up	down	up	down	down
Reduced flow in check valve 6c	down	up	down	up	down	down
Reduced flow in check valve 6d	down	up	down	up	down	down

The complexity of the dynamic modelling approach can be appreciated, for example, by considering the hydraulic circuit of a forging press control system similar to that shown as Figure 1.1. Consider one press concept shown schematically in Figure 1.12. Such a press usually has retracting cylinders in addition to the

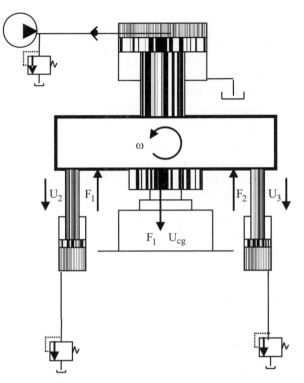

Figure 1.12. Press system schematic in its ready-to-press mode

main press cylinder/s. Also the multiple-pump units are often positioned some distance away from the press with distances that can be more than 100 m.

Such systems have a complex sequence of actions and safety procedures and in the mode shown in the schematic Figure 1.12, the two retracting cylinders are holding the press moving mass at rest via the pressure relief valves. The main pump sets are then switched in to initiate pressing.

For the pressing mode shown, long lines significantly contribute to the system dynamic behaviour and are reflected as the dominant frequency component in the measured press cylinder pressure. Therefore in this application transmission line modelling is a crucial aspect with three lines being in operation.

Press position is usually a smooth characteristic as a consequence of the filtering effect of the large press cylinder volumes. System damping is provided by the pressure relief valves and the large-diameter lines, often with turbulent flow conditions, but the main pressure can have fluctuations during pressing compared with the retracting cylinder pressures, and due to transmission line effects. The main lines do not always have pressure relief valves in operation but the retracting cylinders do have pressure relief valves in operation during pressing. Note that a check valve is often placed in the main press line and this can be designed to improve damping at the cost of a small pressure drop. However, this is not the normal procedure for check valve operation. Any frequency modes from lines and/or fluid volume/moving mass contributions are expected to be low for this type of system.

Computer modelling issues that then have to be addressed are:

- pump flow and controller characteristics
- transmission line dynamics and for large-diameter lines probably with turbulent flow
- very large compressibility flows have to be accommodated
- pressure relief valve, servovalve, directional control valve dynamics and nonlinear flow characteristics, often at very high flow rates where experimental validation is difficult and expensive
- material forging force/displacement properties
- press motion and deformation, force and moment equations
- press control techniques, for example to minimise rotation if off-centre forging is to be done
- cylinder friction forces are difficult to assess for very large cylinders, but they must be included for pressing modes where pressure relief valves are not in operation

The use of dynamic data may be difficult for condition monitoring of such large force systems, but a comparison of pressures, positions, angles, etc. can prove useful when oscillations have decayed and the press is still moving.

Consider, for example a computer model using modelling concepts previously discussed. In this application a pressing force of 22.5 MN is being generated and with rotation angle (Proportional + Integral) control using flow bleed control valves. It is useful to first construct a simulation flow chart to link and thus organise each modelling aspect, and in this example, Figure 1.13 illustrates such a flow chart. Figure 1.14 shows the press performance under normal operation and with a fault at the P + I controller. In this particular example, for this case of controller failure, the dynamic element of performance has not significantly changed. The frequency component is unchanged and there is negligible variation in the press tool position characteristic. However a drift in angle is evident indicating a suspected fault at the controller.

Other fault conditions must of course be studied so that the correct fault prognosis is made. Rapid progress is being made in such computer modelling approaches but more effort is needed to compare simulation models with practical results particularly for very large systems. The next chapter aims to briefly present some of the underlying background theory for modelling components.

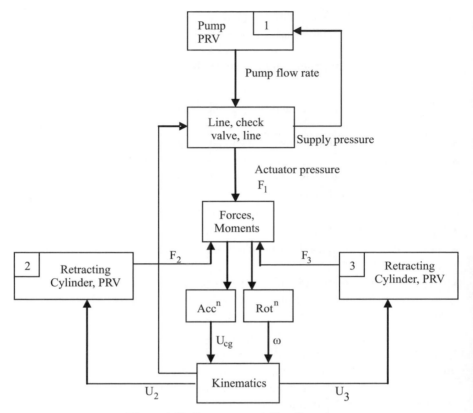

Figure 1.13. Computer modelling flow chart

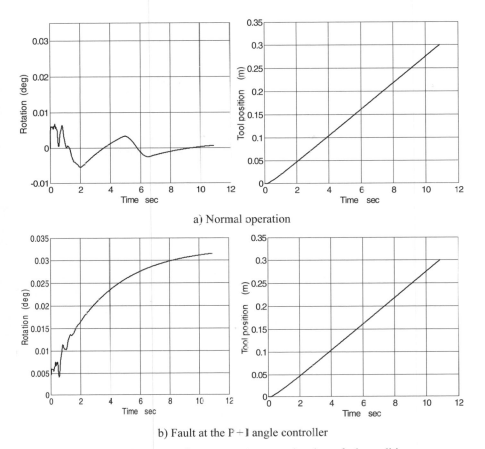

a) Normal operation

b) Fault at the P + I angle controller

Figure 1.14. Press performance under normal and one fault condition

Chapter 2

Modelling and Computer Simulation as an Aid to Understanding Circuit Behaviour

2.1 Introduction

Condition monitoring of hydraulic circuits can perhaps be classified into two practical areas:

(i) Individual component performance assessment, for example wear in a pump with physical indicators such as:

- increased leakage loss
- loss in pump efficiency
- increased noise and/or vibration levels

This may require steady-state torque/speed and pressure/flow measurements or sound level meter/body accelerometer dynamic measurements.

(ii) Hydraulic circuit assessment, either steady-state or dynamic. For example in a cylinder pressure control circuit:

- is pressure control being achieved without pressure oscillations?
- are the demanded pressures being achieved?
- is the control valve functioning correctly?
- are the electronics/transducer working correctly?
- is the product quality within tolerance?

A condition monitoring strategy is clearly significantly helped if the performance of the components and hence the hydraulic circuit can be predicted, particularly in the presence of fault conditions. The ability to run a computer simulation and then match it to the actual circuit performance and deduce faults is perhaps the ultimate goal of the fluid power systems engineer. The simulation may also serve as a useful training package whereby existing or potential faults may be inserted and the change in performance noted.

Unfortunately the vast array of hydraulic components, often with complex internal fluid/mechanical/electronic interactions, makes it almost impossible to generate standard component models for the simulation of circuit dynamic behaviour. Also, it is almost impossible to focus on one particular component since its interaction with other components cannot be neglected in a practical circuit; the modelling of the component of interest inevitably requires the modelling of components connected to it. Much progress has been made in this area but often the models available require experimentally-determined data to set up the computer simulation.

From a component deterioration point of view it is possible in some cases to isolate specific characteristics since they tend to be reflected by easily-attained data, particularly noise (pressure or vibration) or fluid losses. Often a portfolio of cause and effect characteristics can be developed, as discussed in Chapter 1, but in most cases some expert knowledge of the component is required to interpret the data and its implication for component re-design.

Component and system modelling is inevitably concerned with the dynamic behaviour and the fact that control valves, pressure relief valves, motors and cylinders, etc. cannot move instantaneously in response to command signals. Open-loop systems can have oscillatory behaviour, and the reason for this is principally due to the combination of fluid compressibility and moving mass. In practice, steady-state theory is insufficient to explain/predict the response to sudden inputs when a load is demanded to move rapidly from one point to another.

In addition, hydraulic component equations tend to include nonlinear terms making classical mathematical analysis virtually impossible. The principal reason for computer simulation is to gain an understanding of the way complex circuits behave and can be improved by the quick implementation of design changes.

2.2 Steady-state Analysis of Components and Circuits

2.2.1 Pumps and Motors

Considering first the power source, that is positive displacement pumps, there are many types and design variations commercially available. The type chosen usually depends upon the power rating of the application, but their steady-state characteristic has a simple yet general form. Figure 2.1 shows schematic diagrams of the most common types.

Axial piston machines dominate the high-power end of the market and manufacturers are being continually driven to increase pressure levels, power levels, and to further consider the use of water-based fluids. *On board diagnostics* will play an increasing role in the future as operating reliability becomes strategically/financially/safety crucial. This will be aided by the rapid rise in intelligent low-cost microelectronic sensors and information transmission systems.

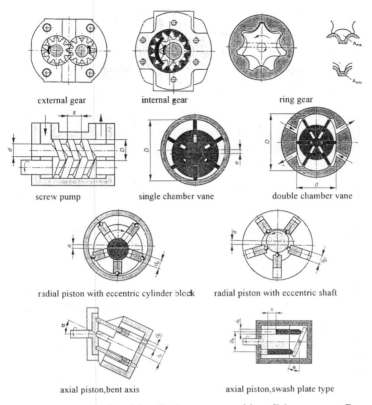

external gear   internal gear   ring gear

screw pump   single chamber vane   double chamber vane

radial piston with eccentric cylinder block   radial piston with eccentric shaft

axial piston,bent axis   axial piston,swash plate type

Figure 2.1. Different types of positive displacement machines [Mannesmann Rexroth, *The Hydraulic Trainer*, Vol 1, 1991]

Considering just an axial piston pump, shown in the schematic Figure 2.2, illustrates the variety of leakage paths that exist due to port plate/body clearance, radial clearance at the pistons, lubricating holes through the slippers.

In practice for any positive displacement pump these flow losses tend to vary linearly with pump pressure, assuming a negligible inlet pressure, and a simple equation may easily be derived to represent the mean flow rate characteristic as shown in Figure 2.3.

From Figure 2.3 the *pump flow* equation is written as follows:

$$Q_p = D_p \, \omega - P/R_p \tag{2.1}$$

The leakage "resistance" R_p for the examples shown is $6 \times 10^{10} \, \text{Nm}^{-2}/\text{m}^3\text{s}^{-1}$ for the axial piston pump and $10 \times 10^{10} \, \text{Nm}^{-2}/\text{m}^3\text{s}^{-1}$ for the smaller capacity vane pump, but both can vary slightly with pump speed. The defining parameter to allow pump selection is its displacement D_p m^3/radian. It then follows that the *torque to drive the pump* is the ideal torque plus that to overcome torque losses as follows:

$$T_p = D_p \, P + T_{\text{losses}} \tag{2.2}$$

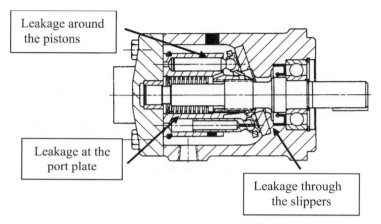

Figure 2.2. Schematic of an axial piston pump

where the losses are usually dominated by a combination of stiction/Coulomb friction and viscous friction terms. These are more important in circuits with motors and linear actuators and will be discussed later. When the pump role is reversed it is then operating as a motor, with some design modifications. *When considering the flow losses for a motor* with both lines pressurised, the flow losses need to be separated. For example a set of experimental results for both flow rates is shown in Figure 2.4, at constant motor speeds for *a motor coupled to a servovalve*.

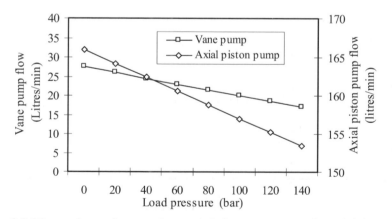

Figure 2.3. Measured mean flow rate characteristic for a vane pump and an axial piston pump

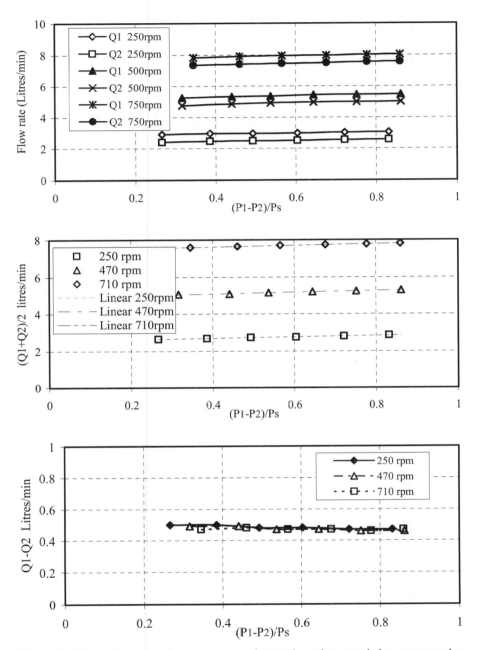

Figure 2.4. Motor flow rates for constant speed operation when coupled to a servovalve [Watton 2005]

Sufficiently accurate flow equations are given by:

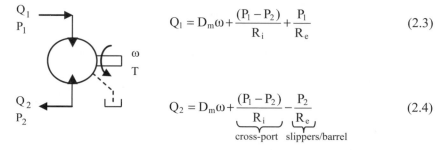

$$Q_1 = D_m \omega + \frac{(P_1 - P_2)}{R_i} + \frac{P_1}{R_e} \tag{2.3}$$

$$Q_2 = D_m \omega + \underbrace{\frac{(P_1 - P_2)}{R_i}}_{\text{cross-port}} - \underbrace{\frac{P_2}{R_e}}_{\text{slippers/barrel}} \tag{2.4}$$

The leakage across the port plate is characterised by a linear leakage resistance R_i and the leakages through the slippers by a linear leakage resistance R_e. Notice that a minimum pressure differential occurs, depending upon the motor speed, and due to the Coulomb/viscous friction effect. The data then leads to a simple method for determining the leakage resistances since from (2.3) and (2.4) the mean flow rate and difference in flow rates are given by

$$Q_{mean} = \frac{(Q_1 + Q_2)}{2} = D_m \omega + (P_1 - P_2) \left[\frac{1}{R_i} + \frac{1}{2R_e} \right] \tag{2.5}$$

$$Q_{difference} = (Q_1 - Q_2) = \frac{(P_1 + P_2)}{R_e} \cong \frac{P_s}{R_e} \tag{2.6}$$

In practice the sum of line pressures is typically $> 0.95 \, P_s$ over a range of typical speed and pressure differentials. Hence the external leakage resistance R_e may be calculated assuming the typical flow difference. Taking the mean line through each set of flow rate data then allows calculation of the motor displacement D_m from (2.4) and also the internal leakage resistance R_i since R_e can be deduced from (2.5).

Considering next the motor torque equation it follows from previous pump theory that, in this example, it may be represented as follows:

$$D_m(P_1 - P_2) = T_m + T_{losses} \tag{2.7}$$

The motor torque losses, T_{losses}, are usually represented by a viscous term proportional to velocity and a nonlinear term indicative of stiction and Coulomb friction. These nonlinear friction terms can have complex forms at very low speeds, and experience dictates their importance for the application of interest. A descriptive form for losses can therefore be written as follows for the purpose of general definition only:

$$T_{losses} = T_{cf} + B_v \, \omega \tag{2.8}$$

where T_{cf} is the stiction/Coulomb friction characteristics which usually has the generalised form shown in Figure 2.5.

The load torque T_m may also contain viscous terms and spring stiffness terms in addition to those caused by the power demand of the application.

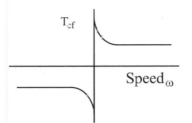

Figure 2.5. Generalised stiction/Coulomb friction characteristics for a motor

2.2.2 Cylinders

A cylinder, or linear actuator, does not usually suffer the inherent leakage characteristic of a motor, or rotary actuator although it is possible for seals to deteriorate and
create a leakage path. The piston and rod seals can provide a stiction/Coulomb friction and viscous friction characteristic, F_{losses}, similar to a motor and will have a form
similar to that shown in Figure 2.5. The flow and force characteristics are as follows:

$$Q_1 = A_1 U \qquad (2.9)$$

$$Q_2 = A_2 U \qquad (2.10)$$

$$P_1 A_1 - P_2 A_2 = F + F_{losses} \qquad (2.11)$$

2.2.3 Leakage Flow and Lift Characteristics of Slippers

Axial piston pumps dominate the high-power end of the market and it is no surprise that piston and slipper leakage flows have received much analytical and
experimental attention, particularly since the 1960s. Some work has brought some
explicit analytical insight into the behaviour of slippers. To demonstrate one analytical area of contribution to component design, consider the piston/slipper assembly shown in Figure 2.6.

Solution of the Reynolds equation for the flow regime across the lands is aided
by the fact that the flow is usually laminar. Also if slipper spin and tilt are considered secondary effects then the radial flow obeys the following simplified equation
for flow across each land:

$$\frac{1}{r^2}\frac{\partial}{\partial \theta}\left(h^3\frac{\partial p}{\partial \theta}\right) + \frac{1}{r}\frac{\partial}{\partial r}\left(rh^3\frac{\partial p}{\partial r}\right) = 6\mu\left(u\cos\theta\frac{\partial h}{\partial r} - \frac{U\sin\theta}{r}\frac{\partial h}{\partial \theta} + \omega\frac{\partial h}{\partial \theta}\right)$$

$$\frac{\partial}{\partial r}\left(\frac{rh^3}{\mu}\frac{\partial p}{\partial r}\right) = 0 \qquad (2.12)$$

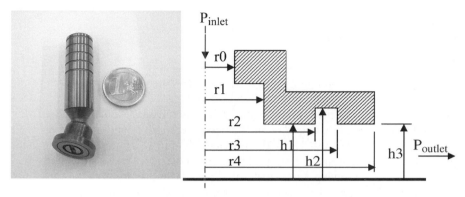

Figure 2.6. Axial piston pump slipper with multiple lands

This equation may be solved by integration across each land and matching inlet and exit boundary conditions at each land. A full computational fluid dynamic analysis has shown that re-circulation effects may be neglected and the explicit solution for flow rate for any n lands, including a groove as a land, may be written in a completely general form as follows (Bergada and Watton, 2004):

$$Q = \frac{\pi}{6\mu} \frac{\left(p_{inlet} - p_{outlet}\right)}{\sum_{i=1}^{i=n} \frac{1}{h_i^3} \ln \frac{r_{(i+1)}}{r_i}} \tag{2.13}$$

The pressure distribution is given by:

(i) for the first land

$$p_1 = p_{inlet} - \frac{\left(p_{inlet} - p_{outlet}\right)}{\sum_{i=1}^{i=n} \frac{1}{h_i^3} \ln \frac{r_{(i+1)}}{r_i}} \left[\frac{1}{h_1^3} \ln\left(\frac{r}{r_1}\right) \right] \tag{2.14}$$

(ii) for the rest of the lands $2 < j < n$

$$p_j = p_{inlet} - \frac{\left(p_{inlet} - p_{outlet}\right)}{\sum_{i=1}^{i=n} \frac{1}{h_i^3} \ln \frac{r_{(i+1)}}{r_i}} \left[\frac{1}{h_j^3} \ln\left(\frac{r}{r_j}\right) + \sum_{j=1}^{j=j-1} \frac{1}{h_j^3} \ln\left(\frac{r_j+1}{r_j}\right) \right] \tag{2.15}$$

Figure 2.7 compares the particular cases of no groove and one groove for an inlet pressure of 160 bar and for the slipper shown in Figure 2.6 and having the following data:

- Central hole diameter $r_o = 0.5$ mm
- Inner land inside radius $r_1 = 5$ mm
- Inner land outside radius $r_2 = 7.43$ mm

- Groove width $= 0.4\,\text{mm}$
- Outside diameter $r_4 = 10.26\,\text{mm}$
- $h_2 = h_1 + 0.4\,\text{mm}$

The pressure distribution is independent of clearance but the flow rate is increased by about 10% when the groove is introduced, as evident from the small increased pressure drop shown in Figure 2.7. Experimental validation is difficult due to the small pressure tapping hole diameter preferred. In addition it is difficult to build an adjustable test rig with sufficient rigidity to avoid distortion under pressure. However, despite these problems the theory has been validated within these constraints allowing an estimate of slipper flow loss to be made providing the clearance is known. This requires an estimate of slipper force balance, and the derived equations may also be applied to this aspect.

The pressure distribution previously shown may be integrated along the slipper to give the hydrostatic lift. This is given by:

$$F_{\text{lift}} = P_{\text{inlet}}\pi\left(r_{(n+1)}^2 - r_0^2\right) - C\pi\, r_n^2 \sum_{i=1}^{i=n}\frac{1}{h_i^3}\ln\left(\frac{r_{(i+1)}}{r_i}\right) + C\pi\sum_{i=1}^{i=n}\frac{1}{h_i^3}\left(\frac{r_{(i+1)}^2 - r_i^2}{2}\right)$$

$$\text{where } C = \frac{P_{\text{inlet}} - P_{\text{outlet}}}{\displaystyle\sum_{i=1}^{i=n}\frac{1}{h_i^3}\ln\left(\frac{r_{(i+1)}}{r_i}\right)} \qquad (2.16)$$

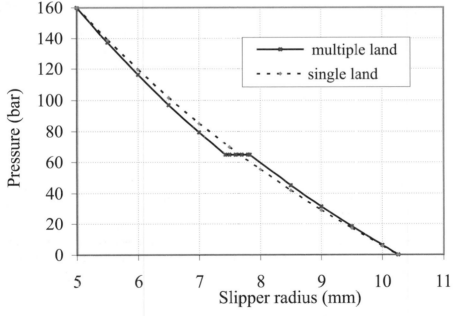

Figure 2.7. Pressure distribution across a slipper with and without a groove at an inlet pressure of 160 bar [Bergada and Watton, 2004]

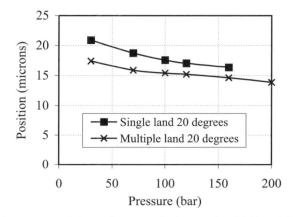

Figure 2.8. Perfect pressure balance for a swash plate angle of 20° and for a slipper with and without a groove [Bergada and Watton, 2004]

The pressure balance, or clamping ratio, is then given by:

$$\alpha = \frac{F_{lift}}{P_{inlet} \, \pi \left(r_p^2 - r_{po}^2 \right)} \tag{2.17}$$

where r_p is the piston diameter and r_{po} is the slipper orifice diameter. The percentage variation from perfect balance is referred to as overbalance, a negative value indicating that hydrodynamic lift is needed and hence slipper tilt. The present theory indicates that clearance is smaller for a grooved slipper but of course flow loss is increased and clearance changes with swash plate angle whatever the design. Figure 2.8 shows the predicted clearance for perfect hydrostatic balance.

2.2.4 Pressure Ripple in Positive Displacement Pumps and Motors

All positive displacement pumps and motors will generate a flow rate ripple superimposed onto the mean flow rate and as a consequence of having a finite number of moving elements, such as gear teeth, vanes, pistons, to generate the flow. This flow ripple, when converted to pressure as dictated by the load characteristic, will always generate vibration and hence noise and must be minimised by good system design in addition to good machine design. Additional noise generated by fluid dynamics within the machine will contribute to the overall noise level producing a wide frequency spectrum beyond the machine "pumping frequency" as dictated by the (machine speed multiplied by the number of moving elements). This is discussed further in Chapter 3.

2.2.5 Flow Restrictors – Flow Rate and Flow Reaction Force

Most fluid components employ fine restrictors and variable areas to control flow rate or pressure drop. By the very nature of fluid power control high pressures are used and hence the net flow areas are small, a flow area of $1\,mm^2$ producing a flow rate of typically 7 litres/min at a pressure drop of 100 bar. Some of the most common flow restrictors are shown in Figure 2.9.

It will be deduced from Figure 2.9 that many control areas vary in a nonlinear manner with displacement. In practice they are all described by a common flow rate equation of the Bernoulli form:

$$Q = C_q a \sqrt{\frac{2\Delta P}{\rho}} \tag{2.18}$$

where a = area, ΔP = pressure differential, C_q = flow coefficient.

Some care is needed in the choice of flow coefficient since it varies with flow rate and experimental validation is usually required. A great deal of work has been done on this topic and the flow coefficient is usually plotted against a defining Reynolds number. For example, Figure 2.10 shows some typical results.

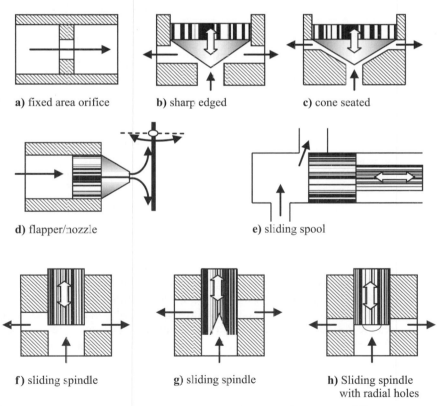

a) fixed area orifice   **b)** sharp edged   **c)** cone seated

d) flapper/nozzle   **e)** sliding spool

f) sliding spindle   **g)** sliding spindle   **h)** Sliding spindle
                                                       with radial holes

Figure 2.9. Some common restrictors used to control pressure and flow rate

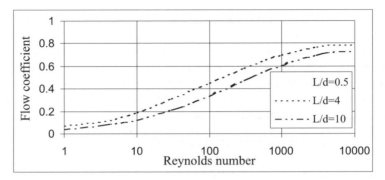

a) fixed area restrictor [Lichtarowicz, Duggins, Markland 1965]

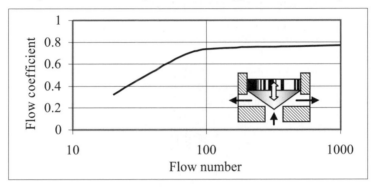

b) 45° poppet valve [McCloy and Martin 1980]

Figure 2.10. Some measured flow coefficient variations for an orifice and a poppet valve

The Reynolds number Re is defined as:

$$Re = \frac{V D_h}{\nu} \tag{2.19}$$

where V = mean velocity and D_h = hydraulic diameter defined as:

$$D_h = \frac{4 \times \text{flow section area}}{\text{flow section perimeter}} \tag{2.20}$$

For a circular orifice the hydraulic diameter is equal to the actual diameter. The problem in practice is determining the mean velocity V since the flow cannot be evaluated using (2.18) in the absence of the flow coefficient. Figure 2.10(a) for an orifice uses the defined Reynolds number while Figure 2.10(b) for a poppet valve uses the flow number which is Reynolds number evaluated with a mean velocity defined as:

$$V = \sqrt{\frac{2\Delta P}{\rho}} \tag{2.21}$$

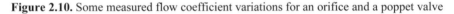

Clearly in practice the very small flow areas result in very large increases in velocity and the resulting momentum change of the fluid must require a force to provide the change. Consequently for a restrictor that utilises a moving element to create a variable area, an equal and opposite force must be experienced by the moving element and this is called the flow reaction force. Consider, for example, the variable restrictor device shown in Figure 2.11. It is assumed that the increased velocity exits at an angle θ to the vertical. Hence the fluid momentum change is given by

$$\uparrow \text{fluid momentum increase} = \rho Q V \cos\theta - \rho Q U$$
$$\approx \rho Q V \cos\theta$$
and using (2.16) $$= 2 C_q^2 \cos\theta \, a \, \Delta P$$

The flow reaction force on the spindle is given by

$$\downarrow \text{Reaction force on spindle} = 2 C_q^2 \cos\theta \, a \, \Delta P \tag{2.22}$$

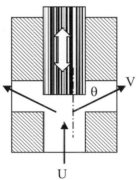

θ V

U

Figure 2.11. Flow reaction force generation

A useful starting point is to consider $2 C_q^2 \cos\theta \approx 1$ although it should be recalled that a more detailed analysis of the flow path may be required if flow reaction force is considered important for a particular design, for example the first stage poppet of a two-stage pressure relief valve. The use of a computational fluid dynamics (CFD) software package is extremely useful in this respect and is illustrated later in the section on servovalves where both flow coefficient and jet angles are considered.

The principle reasons for flow coefficient nonlinear variation with flow rate are:

- Jet angles may not remain constant during operation
- Jet attachment/reattachment can occur
- Flow re-circulation will probably exist
- Cavitation can occur at the downstream side of the restrictor
- Areas may not vary linearly with displacement

2.2.6 Pressure Drop through Pipes

Pressure drop through pipes is an important design consideration particularly for the higher viscosity applications of water-based fluids. Assuming smooth-walled pipes, the well-established equation for pressure drop ΔP is as follows:

$$\Delta P = (4f)\frac{\rho \ell V^2}{2d} \tag{2.23}$$

where 4f is the friction factor, ρ the fluid density, ℓ the pipe length, V is the fluid mean velocity, d the pipe internal diameter. Considering laminar flow occurs for a Reynolds number, typically Re < 2000, and turbulent flow, typically Re > 2000, the friction factor may be determined from either of the following equations:

Laminar flow $\qquad 4f = \dfrac{64}{Re}$

Turbulent flow $\qquad 4f = \dfrac{0.316}{Re^{1/4}} \quad Re = \dfrac{Vd}{v} \tag{2.24}$

For laminar flow, the equations may be combined to give the following pressure drop equation:

$$\Delta P = \frac{128\mu\ell}{\pi d^4}Q \tag{2.25}$$

The term preceding Q in equation (2.25) is often termed the resistance R in fluid power linear circuit theory. Pipe roughness effects are rarely required for hydraulic circuit design, but should this be necessary then the Moody diagram should be used, and may be found in standard Fluid Mechanics books (for example, Nakayama and Boucher, 2002).

2.2.7 Pressure/Flow Characteristics of Directional Valves, Check Valves, Flow Control Valves, Pressure Relief Valves

These common components in a circuit have well-recognised pressure drop characteristics that may easily be measured. However, it is common for manufacturers to provide data which may be used for circuit calculations.

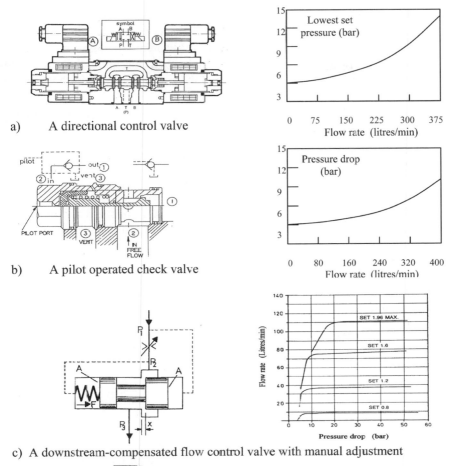

a)   A directional control valve

b)   A pilot operated check valve

c) A downstream-compensated flow control valve with manual adjustment

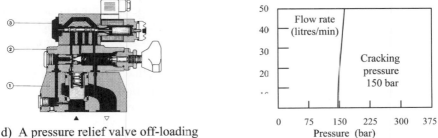

d) A pressure relief valve off-loading

Figure 2.12. Directional, check, flow, and pressure control valve concepts

- **Directional valves** are not intended to be highly restrictive since the objective is to direct flow from the appropriate port to another part of the circuit and with a minimum pressure drop across the valve; the ports are intended to be either fully open or fully closed. They tend to be solenoid-operated, the principle being to open or close the valve as quickly as possible.

- *Check valves* are intended to prevent flow in one direction but to allow it in the opposite direction. The very nature of such a device means that for the ostensibly "free flow" direction a small pressure differential is required to open the valve. The ensuing large flow area when fully open is again intended to create minimum further pressure drops. Check valves may also be pilot-operated to enforce flow in the direction normally closed.
- *Flow control valves* have an internal moving element that compensates for pressure differential changes across them and thus attempts to maintain the constant flow rate set. A minimum pressure differential is however required before the constant flow control regime is achieved and this can be significant in practice.
- *Pressure relief valves* are usually two-stage valves whereby the main stage spool is controlled by a restrictor bridge and first stage poppet valve input element. The poppet valve opening can be manually adjusted or via a solenoid-operated stage, sometimes with position feedback. Pressure relief valves are used either for pressure setting and/or pressure safety protection. They are dynamic components in reality but with a well-defined steady-state pressure/flow characteristic. They may also be off-loaded using a pressure pilot signal or an integral directional control valve.

Schematic diagrams of typical directional, check, flow, pressure relief valves are shown in Figure 2.12 with typical pressure/flow characteristics. Note that there is a very large variation in design for these valves and Figure 2.12 is not definitive by any means.

2.2.8 A Circuit Calculation Example

Consider a lifting system similar to one that has been used to lift concrete bridge sections during a road construction. To produce a sufficient lifting distance, telescopic actuators are often used, but to illustrate circuit calculations a single actuator will be considered as shown in Figure 2.13. Also shown are some common components associated with applications of this type.

To lift the load the directional valve is actuated allowing oil to pass through check valve 3, through the check valve bridge plus flow control valve 1, and through check valve 2(b). Flow is returned from the top of the cylinder through the check valve in parallel with relief valve 4, and then finally across the return port of the directional control valve.

To lower the load the directional valve is actuated in the other direction allowing oil to pass directly through relief valve 4. The setting of this valve is sufficient to open pilot-operated check valves 2(a) and 2(b). The majority of the oil leaving relief valve 4 passes to the top of the cylinder, the residual amount passing through relief valve 5 which must be set to ensure cylinder rod extension. Since pilot-operated check valves 2 are open the cylinder return flow passes through the check valve bridge plus flow control valve 1 and is re-circulated to the top of the

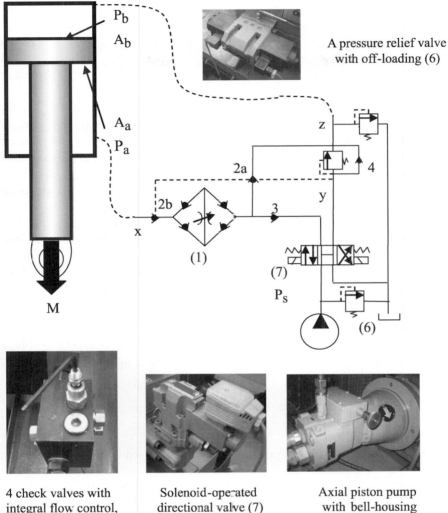

A pressure relief valve
with off-loading (6)

4 check valves with
integral flow control,
cartridge type (1)

Solenoid-operated
directional valve (7)

Axial piston pump
with bell-housing
coupling to motor (8)

Figure 2.13. Bridge section lifting circuit with typical components

cylinder in addition to the flow provided by the pump. In this way flow continuity is maintained thus preventing cavitation.

The use of pilot-operated check valves also ensures that the load does not collapse when the directional control valve is switched to the neutral, and closed, position. It should be noted that this system is a meter-in and meter-out flow control system.

A sufficient pressure drop must exist across the flow control valve to ensure that the correct flow is achieved at the required setting. Both long lines to and from the actuator are of equal length. General information is as follows:

- Load mass $10^5\,\text{kg}$
- Cylinder rod and lifting cradle mass $1500\,\text{kg}$
- Cylinder areas $A_a = 0.05\,\text{m}^2$ $\quad A_b = 0.1\,\text{m}^2$
- Required flow control $30\,\text{litres/min}$
- Friction force at piston and rod seals $F_f = 1000\,\text{N}$
- Selected pump flow rate $40\,\text{litres/min}$

It follows that under correct flow control the cylinder rod should move at a steady-state speed of 10 mm/s.

The flow rate into and out of the top of the cylinder will be double the flow control value due to the area ratio effect, that is 60 litres/min. Hence, lowering the re-circulating flow part of the circuit requires a make-up from the pump of 30 litres/min. A pump flow rate of 40 litres/min has been selected.

(i) Consider lifting and the following pressure drops which have been obtained from manufacturers' data and calculations:

- pressure drop across directional valve supply @ 30 litres/min 4 bar
- pressure drop across directional valve return @ 60 litres/min 10 bar
- pressure drop across each check valve @ 30 litres/min 2 bar
- pressure drop across each check valve @ 60 litres /min 3 bar
- pressure drop across flow control valve @ 30 litres/min 15 bar
- pressure drop down hose @ 30 litres/min 4 bar
- pressure drop down hose @ 60 litres /min 8 bar

To lift a load mass M it is necessary that the following condition be satisfied:

$$P_a A_a > F_f + Mg + P_b A_b \tag{2.26}$$

and considering the supply line:

$$P_s = P_a + \Delta P \tag{2.27}$$

where ΔP is the sum of all the pressure drops at the inlet side. Hence combining these two equations give

$$P_s > \Delta P + \frac{F_f}{A_a} + \frac{Mg}{A_a} + \frac{P_b A_b}{A_a} \tag{2.28}$$

where $\Delta P = 4 + 8 + 15 + 4 = 31\,\text{bar}$ \hfill (2.29)

Now, considering the return section of the circuit, the pressure P_b is equal to the sum of the pressure drops back to tank. This is given by

$P_b =$ pressure drops through the line + check valve 4 + directional valve

$$P_b = 8 + 3 + 10 = 21\,\text{bar} \tag{2.30}$$

pressure relief valve 5 must be set to a value greater than 13 bar \hfill (2.31)

From (2.28):

$P_s > 31 + 0.2 + 199.1 + (21)2 = 272.3$ bar

Supply pressure relief valve 6 setting > 273 bar   (2.32)

Note the error of 73 bar had pressure losses not been included. Note also the effective doubling of the actuator top pressure due to the area ratio of 2.

(ii) Consider lowering with the relief valve 4 setting used to open the pilot-operated check valves. This setting is lower than the main system relief valve at the pump and hence all the pump flow rate passes through relief valve 4. For lowering it is required that:

$P_b A_b + Mg > F_f + P_a A_a$ (2.33)

and for the re-circulating flow:

$P_a = P_b + \Delta P$ (2.34)

where ΔP is the pressure drop around the re-circulation path which now includes both lines, one with a flow rate of 30 litres/min, the other with a flow rate of 60 litres/min.

$\Delta P = 4 + 4 + 15 + 8 = 31$ bar   (2.35)

Therefore from these three equations:

$$P_b \left(\frac{A_b}{A_a} - 1 \right) > \Delta P + \frac{F_f}{A_a} - \frac{M_g}{A_a}$$ (2.36)

$$P_b > 31 + 0.2 - \frac{Mg}{10^5 A_a} \quad \text{bar}$$ (2.37)

The main issue therefore is the case when extending with no load, that is cradle mass only. This then leads to:

$P_b > 28.3$ bar   (2.38)

pressure relief valve 5 must be set to a value greater than 36.3 bar   (2.39)

To determine the relief valve 4 setting, it is usual for pilot-operated check valve to be opened with a pressure 1/3 of the pressure at the check valve. The maximum value will be with full load and with the load suspended at rest and then required to lower. At rest the pressure $P_x = 199$ bar so select a pilot pressure greater than 66 bar. Considering the requirements for both extending and retracting with and without load, a suitable design would be:

- Supply relief valve setting $P_s = 300$ bar
- Relief valve 5 setting $= 50$ bar
- Relief valve 4 setting $= 70$ bar

The setting of relief valve 4 must be such that the pressure drop across it is achievable with a flow rate of 30 litres/min such that relief valve 5 setting is not exceeded. If this happens then relief valve 5 setting is simply increased. The maximum supply power required at the pump drive shaft is therefore 20 kW.

2.3 Electrohydraulic Servovalves

2.3.1 Principles of Operation

Considering a typical servovalve, steady-state ideal flow equations will serve to illustrate the inherent nonlinear characteristic of this crucial hydraulic component used for precision control applications. Figure 2.14 shows a schematic of a common servovalve design.

The electromagnetic first stage produces a torque which may be assumed proportional to current, and the flapper/nozzle stage generates pressures that are applied to each end of the spool. The feedback wire couples the spool to the flapper creating force feedback such that, with a current applied, the flapper rapidly returns to its central position with the spool displaced such that the wire feedback torque balances the electromagnetic torque. The pressure differential generated by the flapper/nozzle stage may be analysed using the schematic shown as Figure 2.15.

In practice the nozzle clearance x_{nm} is significantly smaller than the nozzle diameter and the curtain area dominates the flow characteristic through the nozzle. This type of pressure control device suffers a small inherent flow and power loss. The device is essentially a potential divider that uses the resistance properties of each orifices and nozzle pair. As the flapper is displaced from its null position, one

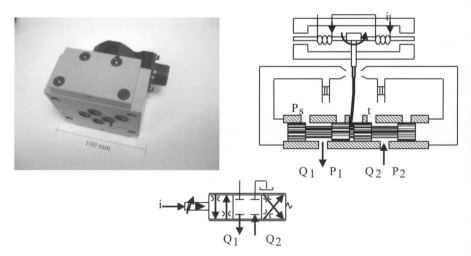

Figure 2.14. Electrohydraulic servovalve

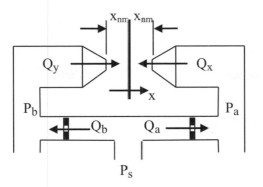

Figure 2.15. Flapper/nozzle stage

pressure increases and the other decreases to create a pressure differential across the spool. Bernoulli-type flow is assumed through each orifice and nozzle and Figure 2.16 shows measured characteristics for a servovalve with matched orifices. In practice each nozzle null pressure is typically half the supply pressure so the working part of Figure 2.16 is towards the right-hand side. The flow equations are then as follows, assuming negligible flow to each end of the spool and a small flapper displacement x:

$$Q_a = C_{qo}a_o\sqrt{\frac{2(P_s - P_a)}{\rho}} \qquad Q_b = C_{qo}a_o\sqrt{\frac{2(P_s - P_b)}{\rho}}$$

$$Q_x = C_{qn}a_{nx}\sqrt{\frac{2P_a}{\rho}} \qquad Q_y = C_{qn}a_{ny}\sqrt{\frac{2P_b}{\rho}} \qquad (2.40)$$

$$a_o = \frac{\pi d_o^2}{4} \qquad a_{nx} = \pi d_n(x_{nm} - x) \qquad a_{ny} = \pi d_n(x_{nm} + x)$$

Re-arranging these equations for the condition $Q_x = Q_a$ and $Q_y = Q_b$ then gives the pressures as follows:

$$\overline{P}_a = \frac{1}{1 + Z(1 - \overline{x})^2} \quad \overline{P}_b = \frac{1}{1 + Z(1 + \overline{x})^2} \quad \overline{P}_a = \frac{P_a}{P_s} \quad \overline{P}_b = \frac{P_b}{P_s}$$

$$\overline{P}_a - \overline{P}_b = \frac{4Z\overline{x}}{(1 + Z(1 - \overline{x})^2)(1 + Z(1 + \overline{x})^2)} \quad \overline{x} = \frac{x}{x_{nm}} \qquad (2.41)$$

The design parameter Z depends on the flow coefficients and the dimensions of each restrictor and may be expressed as follows:

$$Z = 16\left[\frac{C_{qn}}{C_{qo}}\frac{d_n}{d_o}\frac{x_{nm}}{d_o}\right]^2 \qquad (2.42)$$

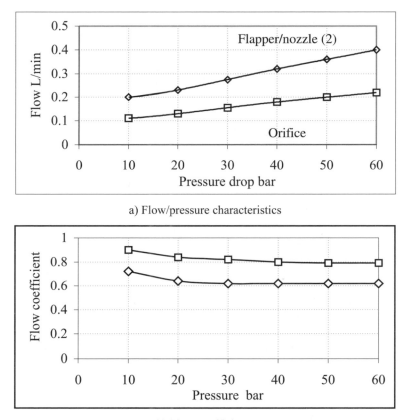

a) Flow/pressure characteristics

b) Flow coefficients

Figure 2.16. Measured flow characteristics for a servovalve orifice/nozzle pair

From Figure 2.16, the appropriate flow coefficients at the higher-pressure end of the working range and other data are:

$C_{qn} \rightarrow 0.62$ $C_{qo} \rightarrow 0.79$

$d_n = 0.47$ mm $x_{nm} = 0.032$ mm $d_o = 0.22$ mm $Z = 0.95$

This is close to the value of $Z \approx 1$ typical of design practice and has been measured for the servovalve under study here. Returning to the pressure characteristic given by (2.24) it can be seen that if $Z = 1$ then each nozzle null pressure is $P_s/2$. Also note from (2.24) that the null gain is given by:

$$\left. \frac{d(\bar{P}_a - \bar{P}_b)}{d\bar{x}} \right]_{\bar{P}_a - \bar{P}_b = 0, \, \bar{x} = 0} = \frac{4Z}{(1+Z)^2} \tag{2.43}$$

This has a maximum value of 1 when $Z = 1$ again suggesting that this value of Z is to be desired in practice. The pressure differential characteristic for $Z = 1$ is shown in Figure 2.17.

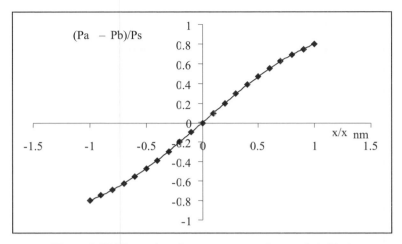

Figure 2.17. Flapper/nozzle stage pressure characteristic $Z = 1$

In practice a drain orifice is used in the flow return line to create a very small drain pressure which helps to stabilise flow characteristic around the flapper. This low-resistance orifice effect is easily introduced into the previous theory, although it has a secondary effect on the pressure gain characteristic shown in Figure 2.17 (Watton, 1987). The principle of operation is then as follows:

- the application of a current to the first stage coils creates an electromagnetic torque which produces flapper rotation and hence a pressure differential as previously described
- this pressure differential acting across the spool creates spool movement and hence a reacting force from the wire
- this reacting force creates a torque at the feedback wire which counteracts the applied electromagnetic torque creating counter rotation of the flapper
- via strategic design of the first and second stages, the flapper returns to its central position with the spool now displaced to a value proportional to applied current.

Erosion of servovalve orifices and flapper/nozzle elements can be a problem in some applications due, for example, to either particle contamination of the oil or water ingress. However, changing technologies are overcoming these issues, such as the use of different materials for the flexure sleeve and sapphire components for the orifices and the ball, which is bonded to the end of the feedback wire. Figure 2.18 shows such developments now available on a selected range by one servovalve manufacturer. Other techniques such as the use of jet pipe technology rather than nozzles to create a pressure control differential are available, with and without the feedback wire.

A complete analysis of a servovalve would need to consider spool flow equations, flow reaction forces and other reaction torques on the flapper together with forces acting on the spool. For steady-state operation most of these issues are

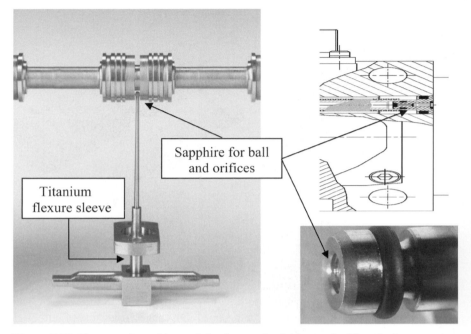

Figure 2.18. The use of sapphire and titanium technologies to significantly improve performance and reliability of servovalves [Star Hydraulics Ltd UK]

secondary, particularly since in practice the servovalve performance is easily obtained or directly measured.

As an example of just one issue, consider an analysis of the flow through the supply and return ports using the FLUENT CFD package. For different spool positions the 3D flow model may be used to compute the flow rate and hence the corresponding flow coefficient using the previously defined theory. In this example the 3D CAD model supplied by the servovalve manufacturer is transposed into the CFD package and the flow grid is meshed within the CAD model. This model is shown in Figure 2.19 together with a typical velocity vector plot for the supply port and computed flow coefficients for both supply and return ports.

Figure 2.19 shows that the trend in the flow coefficient is the same for each port and of an expected form, that is rising as opening increases, reaching a peak and then falling to a constant value, in this case typically $C_q \approx 0.8$. The supply port flow coefficient is slightly higher than the return port flow coefficient for the same spool displacement. An average working value of $C_q \approx 0.85$ is suggested for this servovalve as a working guide. Since $C_{qi} > C_{q2}$ then this implies that the common port pressure for the rated test, ideally $P_s/2$, is actually slightly greater than this value. However, it turns out that the port pressure is only increased by $1 \rightarrow 2$ bar from the ideal value of 35 bar over the spool position range evaluated. A computation of each port flow coefficient therefore requires a very accurate pressure

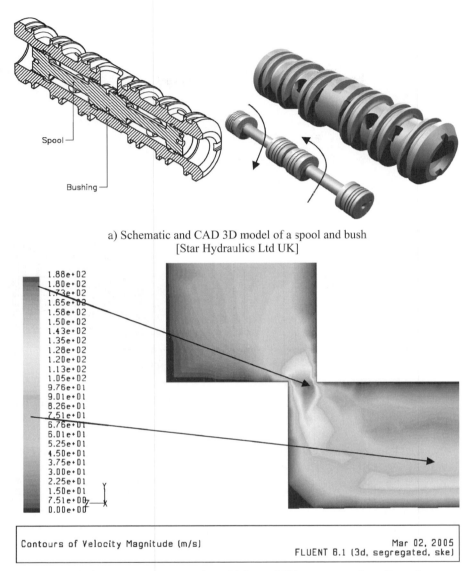

a) Schematic and CAD 3D model of a spool and bush
[Star Hydraulics Ltd UK]

Contours of Velocity Magnitude (m/s)

Mar 02, 2005
FLUENT 6.1 (3d, segregated, ske)

b) Velocity contours at an opening of 0.5mm, 150 bar pressure drop, $P_S \rightarrow A$.
Taken from an axial 2D slice though one of the 4 main ports around the bush.

Figure 2.19. CFD approach to determine the flow coefficient for a typical servovalve spool [Watton and Thorp, Cardiff University, 2005]

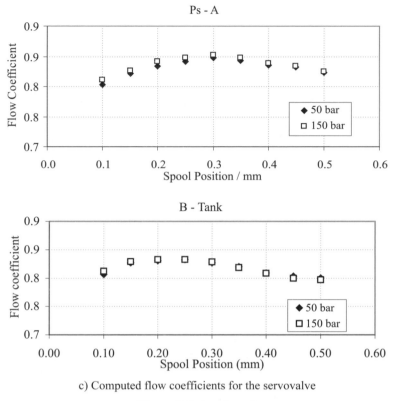

c) Computed flow coefficients for the servovalve

Figure 2.19. (continued)

gauge/transducer, typically with an accuracy of ±0.25 bar, and difficult to achieve. Figure 2.20 shows the variation in jet angle estimated in a number of ways:

- Visually from the velocity vectors using the maximum magnitude, at the main entry and exit sections
- Visually from the velocity vectors using the maximum magnitude, at the remaining peripheral entry and exit section and at a radial angle of 45° from the main entry ports
- By using the CFD-computed axial forces and the momentum change equation to produce an effective single jet angle for the whole spool flow path. The momentum change equation used is as described by equation (2.22).

It is clear from Figure 2.20 that, using the visual approach for the main entry and exit ports, the supply port angle is slightly less that the return port angle. However, given the variation in the data it could be argued that both jet angles are the same, having a typical value of 78°. The less dominant peripheral flow, apart from opening close to zero, shows a decreasing jet angle with increasing spool opening. Using the momentum change equation, and based upon the CFD-computed axial force at the appropriate flow section, gives an almost constant jet angle for both the

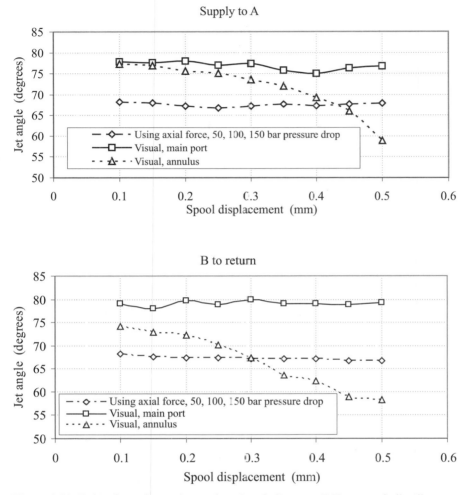

Figure 2.20. Estimation of port jet angles. Spool diameter 7.92 mm, spindle diameter 3.92 mm, ISO 32 Mineral oil

inlet and exit main ports. A typical value of 67° is remarkably close to the von Mises potential flow solution of 69°. Considering either the visual approach or the momentum approach, jet angles at the supply and return ports are predicted to be almost constant and this perhaps partly explains why a servovalve flow characteristic is close to linear as spool displacement is varied. This will be returned to later.

2.3.2 Steady-state Performance Characteristics

Now consider the servovalve flow characteristics. The small port opening areas on either side 'a', vary linearly with spool displacement and the usual Bernoulli flow equations are used by servovalve manufacturers. Following the previous CFD analysis it would seem that the use of a constant flow coefficient may be justified to a very good first approximation.

Hence from Figure 2.21 the flow equations may be written:

$$Q_1 = C_q a \sqrt{\frac{2(P_s - P_1)}{\rho}} \quad Q_2 = C_q a \sqrt{\frac{2P_2}{\rho}} \tag{2.44}$$

The tank (return) pressure is usually neglected in comparison to the line pressures. Also the port opening area 'a' is proportional, to a good first approximation, to spool displacement, which is also proportional to the current applied to the electromagnetic first stage. Servovalve manufacturers also quote *the rated flow* at the valve *rated current* and with a *valve pressure drop of 70 bar*, that is, the total pressure drop across both ports. Consequently the servovalve equations may be re-written in the following form:

$$Q_1 = k_f i \sqrt{P_s - P_1} \quad Q_2 = k_f i \sqrt{P_2} \tag{2.45}$$

valve pressure drop $= (P_s - P_1) + P_2 = P_s - P_{load} = 70$ bar

For this test the ports are connected via a restrictor to generate a load pressure differential and therefore $Q_1 = Q_2$. It is then easy to see from (2.45) that a general flow equation, when combined with the rated flow condition, may be defined as follows:

$$\frac{Q}{Q_{rated}} = \frac{i}{i_{rated}} \sqrt{\frac{P_s - P_{load}}{70}} \tag{2.46}$$

The rated flow is 38 litres/min with series connected coils at 7.5 mA maximum current. There is the usual evidence of a small electrical hysteresis characteristic with a small spool overlap. Whether or not the spool is over-lapped, critically-lapped or under-lapped is dictated by the user and then supplied to order by the manufacturer. Underlap is pursued later.

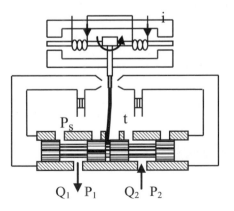

Figure 2.21. Servovalve in operation

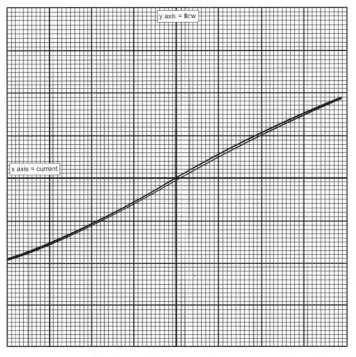

Figure 2.22. Servovalve rated characteristic [Star Hydraulics Ltd UK]

To validate the computed flow coefficients the average data from four new servovalves of the same design and using the total valve pressure drop. Using measured data at the rated conditions the ports A and B are connected and the supply
pressure is set at 70 bar giving ostensibly ε 35 bar pressure drop across each flow
path $P_s \rightarrow A$ and $B \rightarrow$ tank. It is not possible to isolate each orifice for the rated
flow test, due to the absence and accuracy required of the common load ports
pressure measurement, and the flow coefficient across both ports is calculated as
an effective flow coefficient. This is not the case for CFD analysis, the flow out of
port A has an associated flow coefficient C_{q1} and the flow into port B has an associated flow coefficient C_{q2}, and shown in Figure 2.19. Therefore, an effective flow
coefficient C_q may be determined as follows:

computed from CFD analysis $\quad Q = \dfrac{C_{q1}C_{q2}}{\sqrt{C_{q1}^2 + C_{q2}^2}} a \sqrt{\dfrac{2P_s}{\rho}}$

rated valve condition tests $\qquad Q = C_q a \sqrt{\dfrac{P_s}{\rho}} \quad P_s = 70 \text{ bar}$ 　　　(2.47)

$$C_q = \dfrac{C_{q1}C_{q2}}{\sqrt{C_{q1}^2 + C_{q2}^2}}$$

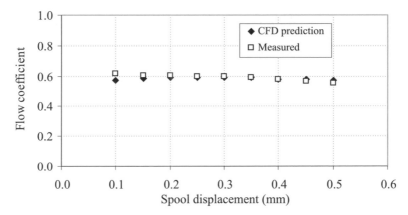

Figure 2.23. Comparison between CFD and measured effective flow coefficient for a servovalve

This comparison is shown in Figure 2.23 where it can be seen that the CFD prediction using individual flow coefficients compares well with that experimentally determined assuming an effective flow coefficient. There is a general measured downward trend in effective flow coefficient with increased spool opening, but the general conclusions reflect both the effectiveness of the CFD approach and the flow linearity of the servovalve, particularly at larger spool displacements. The CFD predictions of course become more computationally difficult as the spool opening is reduced due to increasingly difficult meshing problems at the flow restriction volume.

2.3.3 Spool Underlap

Before closed-loop control is pursued it is useful to look at spool underlap. Spool underlap contributes to damping for many control systems applications and is often specified by the user and/or the manufacturer. Assume spool symmetrical underlap and consider Figure 2.24. The machined underlap on each

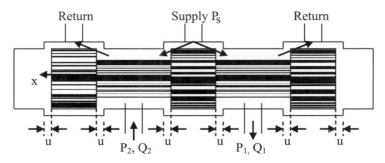

Figure 2.24. Symmetrically lapped 5-way spool valve

spool land is "u" and the servovalve current to just move the spool this distance is designated i_u.

Considering spool displacement to the left and within the underlap region gives:

$$Q_1 = k_f(i_u + i)\sqrt{P_s - P_1} - k_f(i_u - i)\sqrt{P_1}$$
$$Q_2 = k_f(i_u + i)\sqrt{P_2} - k_f(i_u - i)\sqrt{P_s - P_2}$$

(2.48)

These equations are only valid for $-i_u \leq i \leq i_u$ and the equations then revert to just the first term on the right-hand side outside the underlap region. Figure 2.25 shows a measured characteristic for a servovalve supplied with an ostensibly

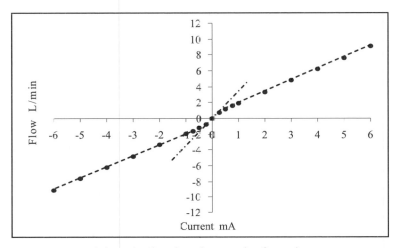

a) A randomly-selected servovalve for testing

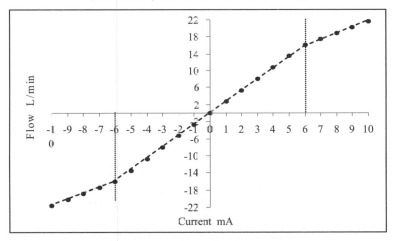

b) A servovalve having a design underlap equivalent to 6mA

Figure 2.25. Flow rate/current characteristics for two different servovalves, supply pressure 100 bar, 40°C, zero load pressure

critically-lapped spool together with one with a deliberately introduced symmetrical underlap equivalent to 6 mA.

It can be seen from Figure 2.25(a) that the flow gain is greatest at the null condition and then falls and becomes almost constant at larger applied currents. In fact the null gain is again about twice that away from the null condition. A small underlap exists and was unknown before the test. Further studies reveal that this underlap is equivalent to typically ±0.27 mA. It can be clearly seen from Figure 2.25(b) that the flow gain within the underlap region is twice that outside the underlap region. Measuring the servovalve flow characteristics with the ports connected via a load restrictor then means that equations (2.48) become:

$$Q = k_f \left(i_u + i\right)\sqrt{\frac{P_s - P_{load}}{2}} - k_f \left(i_u - i\right)\sqrt{\frac{P_s - P_{load}}{2}} \tag{2.49}$$

$$Q = 0 \quad \text{when} \quad \overline{P}_{load} = \frac{2\,\overline{i}}{\left(1 + \overline{i}^2\right)} \quad \overline{P}_{load} = \frac{P_{load}}{P_s} \quad \overline{i} = \frac{i}{i_u} \tag{2.50}$$

2.3.4 Spool Valve Linearised Coefficients

The approach, also known as a small signal analysis, considers small movements about the steady-state condition. The method is commonly applied to servovalve-controlled systems whereby the nonlinear flow characteristics are linearised and then combined with the other system dynamic terms. This enables a linearised transfer function to be developed which may then be used for conventional linear systems analysis and design. Such an approach can give valuable insight into the way parameters affect system performance.

$$Q_1 = k_f i\sqrt{P_s - P_1}$$
$$Q_2 = k_f i\sqrt{P_2} \tag{2.51}$$

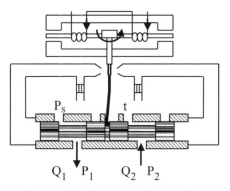

Figure 2.26. Servovalve in operation

Consider first the servovalve flow equations, Figure 2.26, and consider variations about an operating condition i_{ss}, P_{1ss}, P_{2ss}, Q_{1ss}, Q_{2ss}, by using just the first, linear, term of the Taylor series expansion for a nonlinear function. Consequently small changes in each parameter leads to:

$$\delta Q_1 = k_{i1}\delta i - k_{p1}\delta P_1$$
$$\delta Q_2 = k_{i2}\delta i - k_{p2}\delta P_2$$

(2.52)

$$k_{i1} = \frac{\partial Q_1}{\partial i} = k_f\sqrt{P_s - P_{1ss}} = \frac{Q_{1ss}}{i_{ss}} \qquad \text{flow gain}$$

$$k_{i2} = \frac{\partial Q_2}{\partial i} = k_f\sqrt{P_{2ss}} = \frac{Q_{2ss}}{i_{ss}} \qquad \text{flow gain}$$

(2.53)

$$k_{p1} = -\frac{\partial Q_1}{\partial P_1} = \frac{k_f i_{ss}}{2\sqrt{P_s - P_{1ss}}} = \frac{Q_{ss}}{2(P_s - P_{ss})} \qquad \text{pressure coefficient}$$

$$k_{p2} = \frac{\partial Q_2}{\partial P_2} = \frac{k_f i_{ss}}{2\sqrt{P_{2ss}}} = \frac{Q_{ss}}{2P_{2ss}} \qquad \text{pressure coefficient}$$

The pressure sensitivity is defined as

$$k_{ip} = \frac{\partial P}{\partial i} = -\frac{\partial Q}{\partial i}\Big/\frac{\partial Q}{\partial P}$$
$$= \frac{k_i}{k_p} = \frac{2(P_s - P_{1ss})}{i_{ss}} \quad \text{or} \quad \frac{2P_{2ss}}{i_{ss}}$$

(2.54)

It can therefore be seen that at the null condition, $P_{1ss} = P_{2ss} = P_s/2$ and $i_{ss} = 0$, for a critically-lapped spool:

Flow gain $\qquad\qquad\qquad k_{i1} = k_{i2} = k_i = k_f\sqrt{P_s/2}$

Pressure coefficient $\quad k_{p1} = k_{p2} = k_p = 0$ (2.55)

Pressure sensitivity $\quad k_{ip1} = k_{ip2} = k_{ip} = \infty$

Alternatively, using the flow equation expressed in terms of load pressure differential:

$$Q = k_f i\sqrt{\frac{P_s - P_{load}}{2}} \qquad \delta Q = k_i\delta i - k_p\delta P_{load}$$

(2.56)

$$k_i = k_f\sqrt{\frac{P_s - P_{loadss}}{2}} \qquad k_p = \frac{k_f i_{ss}}{2\sqrt{2}\sqrt{P_s - P_{loadss}}}$$

At the null condition, $P_{loadss} = 0$ and $i_{ss} = 0$, the linearised coefficients are:

Flow gain $\qquad\qquad\qquad k_i = k_f\sqrt{P_s/2}$

Pressure coefficient $\quad k_p = 0$ (2.57)

Pressure sensitivity $\quad k_{ip} = \infty$

Since the pressure coefficient term contributes to damping it can be said that a critically-lapped spool offers no contribution to system damping at the zero current condition. However, this is not the case if an under-lapped spool is considered. Recalling (2.48) **for a symmetrically under-lapped spool**, the linearised coefficients may also be derived:

$$Q_1 = k_f(i_u + i)\sqrt{P_s - P_1} - k_f(i_u - i)\sqrt{P_1}$$
$$Q_2 = k_f(i_u + i)\sqrt{P_2} - k_f(i_u - i)\sqrt{P_s - P_2} \tag{2.58}$$

Linearising these equations then gives:

$$\delta Q_1 = k_{i1}\delta i - k_{p1}\delta P_1$$
$$\delta Q_2 = k_{i2}\delta i - k_{p2}\delta P_2$$

$$k_{i1} = \frac{\partial Q_1}{\partial i} = k_f\sqrt{P_s - P_{1ss}} + k_f\sqrt{P_{1ss}} \qquad \text{Flow gain}$$

$$k_{i2} = \frac{\partial Q_2}{\partial i} = k_f\sqrt{P_{2ss}} + k_f\sqrt{P_s - P_{2ss}} \qquad \text{Flow gain} \tag{2.59}$$

$$k_{p1} = -\frac{\partial Q_1}{\partial P_1} = \frac{k_f(i_u + i)}{2\sqrt{P_s - P_{1ss}}} + \frac{k_f(i_u - i)}{2\sqrt{P_{1ss}}} \qquad \text{Pressure coefficient}$$

$$k_{p2} = \frac{\partial Q_2}{\partial P_2} = \frac{k_f(i_u + i)}{2\sqrt{P_{2ss}}} + \frac{k_f(i_u - i)}{2\sqrt{P_s - P_{2ss}}} \qquad \text{Pressure coefficient}$$

Therefore for the null condition, $P_{loadss} = 0$ and $i_{ss} = 0$, it can be seen that the linearised coefficients become:

Flow gain $\qquad\qquad k_{i1} = k_{i2} = k_i = 2k_f\sqrt{P_s/2}$

Pressure coefficient $\quad k_{p1} = k_{p2} = k_p = \dfrac{k_f i_u}{\sqrt{P_s/2}} \tag{2.60}$

Pressure sensitivity $\quad k_{ip1} = k_{ip2} = k_{ip} = \dfrac{P_s}{i_u}$

Comparing (2.60) with (2.57) for a critically-lapped spool it can be seen that for the under-lapped spool:

- the flow gain has doubled
- the pressure coefficient is finite, hence damping at null is provided
- the pressure sensitivity is now finite

Alternatively using (2.49) for the case with flow expressed in terms of load pressure differential gives:

$$Q = k_f(i_u + i)\sqrt{\frac{P_s - P_{load}}{2}} - k_f(i_u - i)\sqrt{\frac{P_s + P_{load}}{2}} \tag{2.61}$$

$$\delta Q = k_i\delta i - k_p\delta P_{load}$$

Flow gain
$$k_i = k_f \sqrt{\frac{P_s - P_{loadss}}{2}} + k_f \sqrt{\frac{P_s + P_{loadss}}{2}}$$

Pressure coefficient $k_p = \dfrac{k_f(i_u + i_{ss})}{4\sqrt{\dfrac{P_s - P_{loacss}}{2}}} + \dfrac{k_f(i_u - i_{ss})}{4\sqrt{\dfrac{P_s + P_{loadss}}{2}}}$ (2.62)

Therefore at the null condition, $P_{loadss} = 0$ and $i_{ss} = 0$, the linearised coefficients are:

Flow gain $k_i = 2k_f \sqrt{P_s/2}$

Pressure coefficient $k_p = k_f i_u / 2\sqrt{P_s/2}$ (2.63)

Pressure sensitivity $k_{ip} = \dfrac{2P_s}{i_u}$

Note the changes from (2.60), which defines the linearised coefficients in a different way. The flow gain is identical, the pressure coefficient is halved and the pressure sensitivity is doubled. From (2.61) it is a simple matter to show that *for blocked load conditions:*

$$\bar{P}_{1ss} = \frac{(1 + \bar{i})^2}{2(1 + \bar{i}^2)} \quad \bar{P}_{2ss} = \frac{(1 - \bar{i})^2}{2(1 + \bar{i}^2)} \quad \bar{P}_{loadss} = \frac{2\bar{i}}{(1 + \bar{i}^2)}$$

$$\bar{P}_{1ss} = \frac{P_{1ss}}{P_s} \quad \bar{P}_{2ss} = \frac{P_{2ss}}{P_s} \quad \bar{P}_{loadss} = \frac{P_{loadss}}{P_s} \quad \bar{i} = \frac{i}{i_u}$$ (2.64)

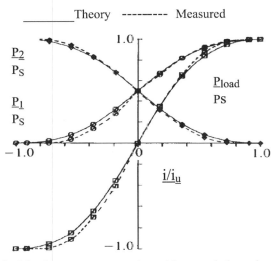

Figure 2.27. Blocked load test on a servovalve with an underlapped spool, the underlap equivalent to 6 mA [Watton and Al-Baldawi 1991]

Hence by connecting the ports together and stroking the servovalve through its underlap current range and measuring the resulting differential, validation of the underlap-equivalent current, i_u, can be done since the the the slope at the origin is the pressure sensitivity. A measured characteristic for a sevovalve having a deliberately introduced underlap is shown in Figure 2.27. It can be deduced that the slope at the origin is ≈ 2.0 as predicted by theory.

2.3.5 Servovalve Dynamic Response

In control applications where servovalve dynamics are thought to be significant, some form of dynamic model must be considered. Higher-order dynamic effects are dominated by spool motion but generally the dynamic performance of a suitably-selected servovalve is superior to the complete system in which it is used. There has been much research in this area, but manufacturers' data is usually provided either in transient response form or the much more common frequency response form. For example, a second-order approximation is often used relating servovalve spool movement, x, to applied current, i:

$$G(s) = \frac{x(s)}{i(s)} = \frac{K}{\left(1 + \frac{2\zeta}{\omega_n}s + \frac{s^2}{\omega_n^2}\right)} \tag{2.65}$$

where K, the gain, ζ the damping ratio, and the undamped natural frequency ω_n are selected from either the manufacturer's data or from other laboratory measurements. Typically a spool movement of 0.5 mm is achieved at rated current. It is also known that for current amplifier-driven servovalves the current response time is of the order of 1 ms resulting in a similarly fast flapper rotation response and hence pressure differential response across the spool.

For example, consider frequency response data from one servovalve manufacturer, Figure 2.28. The frequency response of all servovalves depends upon the load chosen and the amplitude of the applied sinusoidal current, expressed as a percentage of the rated current, as may be seen from Figure 2.28. The deterioration at higher amplitudes is a measure of the nonlinear flow characteristic across the ports combined with the spool dynamic motion characteristic. In fact an approximation for the frequency response, in this case for 25% input signal, suggests a second-order transfer function having a damping ratio of typically $\zeta = 0.55$ and an undamped natural frequency of typically 230 Hz, valid up to about 250 Hz. This is representative of a very fast-acting servovalve.

For more detailed system studies a third-order transfer function could be determined for very fast-acting servovalves. Frequency response, however, will deteriorate for servovalves with high rated flow rates, very high flow rates often requiring two hydraulic stages.

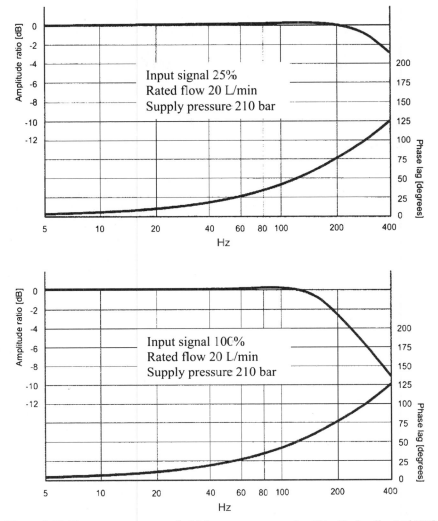

Figure 2.28. Frequency response of a high-response servovalve [Star Hydraulics Ltd UK]

2.4 Steady-state Control of a Servovalve/Motor Drive

2.4.1 The Basic Circuit

Consider the formation of a basic control system by coupling a servovalve to a motor as shown in Figure 2.29. A constant load resisting torque T is supplied by the load pump with integral pressure relief valve and boost pump for flow make-up.

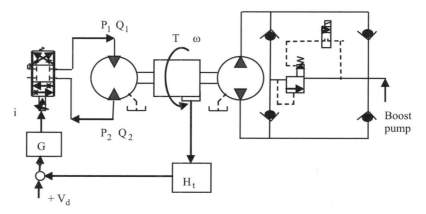

Figure 2.29. Basic servovalve/motor drive with loading circuit

Following earlier considerations of servovalve and motor steady-state characteristics in Section 2.2, we arrive at the basic equations as follows:

$$Q_1 = k_f i \sqrt{P_s - P_1} = D_m \omega + \frac{(P_1 - P_2)}{R_i} + \frac{P_1}{R_e}$$

$$Q_2 = k_f i \sqrt{P_2} = D_m \omega + \frac{(P_1 - P_2)}{R_i} - \frac{P_2}{R_e} \tag{2.66}$$

$$D_m (P_1 - P_2) = T + T_{losses} + B_v \omega$$

$$i = Ga (V_d - H_t \omega)$$

2.4.2 Open-loop Behaviour with Losses

It is not possible to present an explicit solution set of equations when losses are taken into account. However, an insight into the behaviour may be gained following the constant speed experimental data presented earlier and by noting that the sum of line pressures is approximately constant, in this study $\approx 0.96\,P_s$ for a range of speeds and loads. To determine the nature of the open-loop and closed-loop characteristics the assumption that $P_1 + P_2 \approx P_s$ may be used since there is a negligible difference in the mean flow rate calculation compared with the exact values over a wide load pressure range, typically up to 90% of the supply pressure. Consequently the flow equations derived earlier now become:

$$Q_1 \cong Q_2 \approx k_f i \sqrt{\frac{P_s - P_{load}}{2}} \tag{2.67}$$

The originally-decoupled pressures P_1 and P_2 may now be expressed in terms of the load pressure resulting in the following explicit equations:

Speed:

$$\overline{\omega} = \sqrt{1 - \overline{P}_{load}} - \alpha \overline{P}_{load} \tag{2.68}$$

Power transferred to the motor:

$$\overline{W}_m = \overline{P}_{load} \sqrt{1 - \overline{P}_{load}} + \varepsilon \tag{2.69}$$

Efficiency:

$$\eta = \left(\overline{P}_{load} - \overline{T}_{losses} \right) \left(1 - \frac{\alpha \overline{P}_{load}}{\sqrt{1 - \overline{P}_{load}}} \right) \tag{2.70}$$

$$\overline{\omega} = \frac{\omega}{\omega(0)} \qquad \overline{T}_{losses} = \frac{T_{losses}}{P_s D_m} \qquad \overline{P}_{load} = \frac{P_{load}}{P_s} \tag{2.71}$$

$$\overline{W}_m = \frac{W_m}{P_s D_m \omega(0)} \qquad \alpha = \frac{P_s}{D_m \omega(0) R_m} \qquad \varepsilon = \frac{P_s}{2 D_m \omega(0) R_e} \tag{2.72}$$

$$\frac{1}{R_m} = \frac{1}{R_i} + \frac{1}{2R_e} \tag{2.73}$$

It will be seen from (2.68) and (2.70) that both the speed and efficiency fall to zero at a load pressure just below supply pressure. Figure 2.30 shows some open-loop measured and theoretical characteristics for two no-load speeds of 234 rpm and 903 rpm resulting from α values of 0.153 and 0.04.

Data applicable to these measurements are:

- Shell Tellus ISO 32 mineral oil @40°C
- $D_m = 1.68 \times 10^{-6}\,\text{m}^3/\text{rad}$, $P_s = 100\,\text{bar}$
- Tacho gain $H_t = 0.00955\,\text{V/rad s}^{-1}$
- Motor and load $T_{losses} = 2\,\text{Nm}$ @ 234 rpm and 3 Nm @903 rpm.

This gives $\overline{T}_f = 0.12$ or 0.18. These values barely change with load pressure at the two test conditions.

Therefore from the open-loop data:

- $R_e = 1.28 \times 10^{12}\,\text{Nm}^{-2}/\text{m}^3\text{s}^{-1}$
- $R_i = 4.2 \times 10^{12}\,\text{Nm}^{-2}/\text{m}^3\text{s}^{-1}$
- $R_m = 1.59 \times 10^{12}\,\text{Nm}^{-2}/\text{m}^3\text{s}^{-1}$

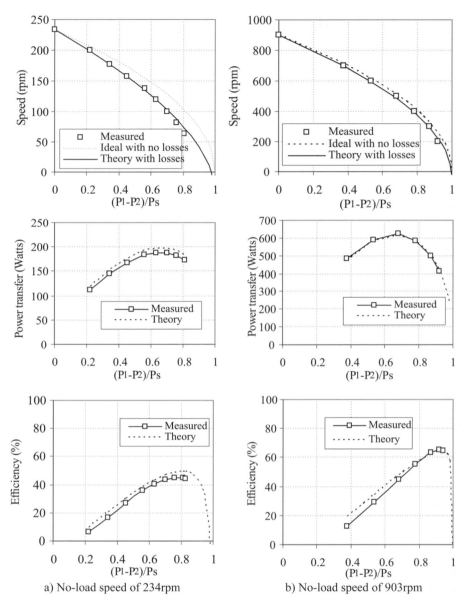

Figure 2.30. Measured and theoretical open-loop characteristics of the servovalve/motor drive

It is deduced that $\alpha = 35.7/N(0)$, $\varepsilon = 22.2/N(0)$ where $N(0)$ is the no-load speed, rpm.

Considering then the conditions for *maximum power transfer shows that this occurs at a load pressure* $(P_1 - P_2) = (2/3) P_s$ *and independent of the motor losses.*

The *maximum efficiency* occurs at the following condition:

$$\alpha = \frac{2(1-\overline{P}_{load})^{3/2}}{\overline{P}_{load}(4-3\overline{P}_{load})-\overline{T}_{losses}(2-\overline{P}_{load})} \tag{2.74}$$

$$\text{maximum } \eta = \frac{(\overline{P}_{load}-\overline{T}_{losses})^2(2-\overline{P}_{load})}{\overline{P}_{load}(4-3\overline{P}_{load})-\overline{T}_{losses}(2-\overline{P}_{load})} \tag{2.75}$$

The theoretical development leads to sufficiently-accurate predictions with generally a small theoretical over-estimate in most cases, particularly at lower speeds where the assumption of a constant sum of line pressures begins to depart from the conditions leading to the assumption. The comparisons become satisfactory as the load pressure is increased towards realistic values that will probably be desired by virtue of the increasing demand to improve operating efficiency. Nevertheless, the prediction of optimum conditions is remarkably good for both power transfer and efficiency, and at both speeds. Because of the small values of α considered, the motor speed falls to zero at load pressures very close to supply pressure. However, the effect is well-demonstrated for the lower no-load speeds.

2.4.3 Closed-loop Behaviour with Losses

Now consider the case with speed feedback and with basic error proportional control since it is well known that this leads to an improved speed/load pressure characteristic. Equations (2.68), (2.69), (2.70) are now modified as follows:

Speed:

$$\overline{\omega} = \frac{(1+K)\sqrt{1-\overline{P}_{load}} - \alpha\overline{P}_{load}}{1+K\sqrt{1-\overline{P}_{load}}} \tag{2.76}$$

Power transferred to the motor:

$$\overline{W}_m = \overline{P}_{load}\left[\frac{(1+K)\sqrt{1-\overline{P}_{load}} - \alpha\overline{P}_{load}}{1+K\sqrt{1-\overline{P}_{load}}}\right] + \alpha\overline{P}_{load}^2 + \varepsilon \tag{2.77}$$

Efficiency:

$$\eta = \frac{(\overline{P}_{load}-\overline{T}_{losses})\left(1-\dfrac{\alpha\overline{P}_{load}}{(1+K)\sqrt{1-\overline{P}_{load}}}\right)}{1+\dfrac{\alpha K\overline{P}_{load}}{(1+K)}} \tag{2.78}$$

$$K = \frac{k_f GaH_t}{D_m}\sqrt{\frac{P_s}{2}} \tag{2.79}$$

It can be seen from the speed equation (2.76) that switching in speed feedback improves the speed droop even in the presence of a severe loss condition as represented by high values of α. Considering the condition when the speed falls to zero gives:

$$\frac{\sqrt{1-\overline{P}_{load}}}{\overline{P}_{load}} = \frac{\alpha}{(1+K)} \tag{2.80}$$

Compared with the open-loop case at the same no-load speed, increasing the gain K is equivalent to reducing the loss coefficient α which increases the load pressure to give zero speed and gives an improvement that 'appears to partially nullify the leakage effect'. It can be seen from equation (2.77) that the total flow loss/load pressure effect on power transfer is now not removed due to the decoupling effect of speed feedback. This suggests that the optimum power is now dependent on the total flow loss term α rather than just the external leakage loss term ε. Considering this ***maximum power transfer*** condition now gives the implicit solution:

$$\alpha = \left[\frac{1+K}{K}\right]\left[\frac{2K(1-\overline{P}_{load})^{3/2} - (3\overline{P}_{load} - 2)}{\overline{P}_{load}(5 - 4\overline{P}_{load}) - 4K\overline{P}_{load}(1-\overline{P}_{load})^{3/2}}\right] \tag{2.81}$$

For the no-loss case, $\alpha = 0$, equation (2.81) does validate the previously-known solution:

$$\overline{P}_{load} = \frac{2}{3} + \frac{2}{3}K\left(1-\overline{P}_{load}\right)^{3/2} \tag{2.82}$$

The condition for ***maximum efficiency*** is more complicated compared with the open-loop case, the result being:

$$\alpha^2 \overline{P}_{load}^2 \left[\sqrt{1-\overline{P}_{load}} - \frac{K\left(\overline{P}_{load} - \overline{T}_{losses}\right)}{2}\right]$$

$$-\alpha(1+K)\left[2\overline{P}_{load}\left(1-\overline{P}_{load}\right) + \left(\overline{P}_{load} - \overline{T}_{losses}\right)\left(K\left(1-\overline{P}_{load}\right)^{3/2} + 1 - \frac{\overline{P}_{load}}{2}\right)\right] \tag{2.83}$$

$$+(1+K)^2(1-\overline{P}_{load})^{3/2} = 0$$

Figure 2.31 shows some measured and computed characteristics for the open-loop and two closed-loop conditions. Due to the minimum pressure differential available, the no-load speed is difficult to assess for each test condition. Therefore a reference condition at a common speed of 500 rpm and a common load pressure of 30 bar has been used for speed resulting in slightly different no-load speeds and hence slightly different values of α. The flattening effect on the characteristic can be clearly validated over the lower load pressure range, but it should be mentioned that the higher gain K = 2.6 resulted in noticeable yet small oscillations in pressure indicating that the condition for closed-loop instability would probably be approached for a further doubling of gain.

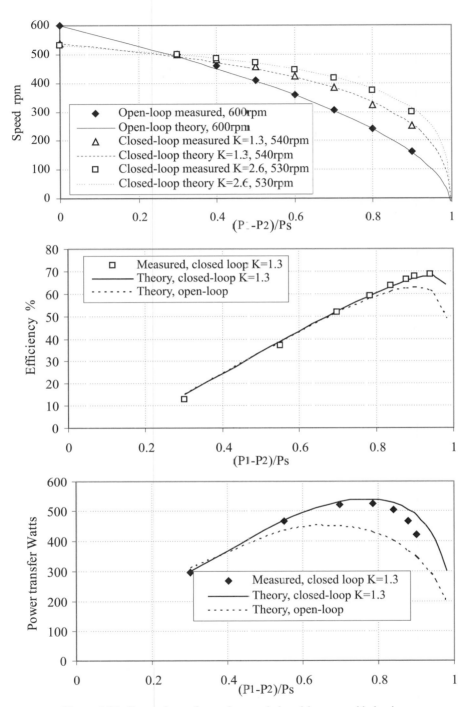

Figure 2.31. Comparison of open-loop and closed-loop speed behaviour

2.5 Steady-state Motion of a Servovalve/Linear Actuator Drive

2.5.1 The Basic Circuit

Now consider the formation of a basic control system by coupling a servovalve to a double acting single rod cylinder as shown in Figure 2.32. The load mass is M and a constant load resisting force F is also included.

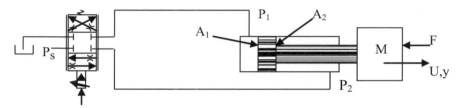

Figure 2.32. Basic servovalve and actuator coupled by short lines

Applying the previously developed system flow continuity and load force equations then gives:

2.5.2 The Extending Case

$$Q_1 = k_f i\sqrt{P_s - P_1} = A_1 U$$
$$Q_2 = k_f i\sqrt{P_2} = A_2 U \tag{2.84}$$
$$P_1 A_1 - P_2 A_2 = F + F_{cf} + B_v U$$

The velocity of the load is U and the displacement of the load is y. It can be seen that for steady-state conditions, and neglecting minor losses for the usual case of a high load force application for cylinder drives:

$$P_1 = \frac{P_s + \gamma^2 P_{load}}{(1+\gamma^3)} \quad P_2 = \frac{\gamma P_s - P_{load}}{(1+\gamma^3)}$$

$$U_{ext} = \frac{k_f i}{A_2}\sqrt{\frac{\gamma P_s - P_{load}}{(1+\gamma^3)}} \tag{2.85}$$

The area ratio $\gamma = \dfrac{A_1}{A_2} \quad P_{load} = \dfrac{F}{A_2}$ $\tag{2.86}$

It is deduced that for the extending case, motion ceases when $F = P_s A_1$ and the pressures are $P_1 = P_s$ and $P_2 = 0$.

2.5.3 The Retracting Case

Assuming positive flow rates but now with Q_1 out of the cylinder, Q_2 into the cylinder and positive acceleration but reversed in direction:

$$Q_1 = k_f i \sqrt{P_1} = A_1 U$$
$$Q_2 = k_f i \sqrt{P_s - P_2} = A_2 U \tag{2.87}$$
$$P_1 A_1 - P_2 A_2 = F + F_{fc} + B_v U$$

Now we see that for steady-state conditions, and neglecting minor losses:

$$P_1 = \frac{\gamma^2 P_s + \gamma^2 P_{load}}{\left(1+\gamma^3\right)} \quad P_2 = \frac{\gamma^3 P_s - P_{load}}{\left(1+\gamma^3\right)}$$

$$U_{ret} = \frac{k_f i}{A_2} \sqrt{\frac{P_s + P_{load}}{\left(1+\gamma^3\right)}} \tag{2.88}$$

It is deduced that for the retracting case, motion ceases when $F = -P_s A_2$ and the pressures are $P_1 = 0$ and $P_2 = P_s$. Hence the load force must be in the opposite direction, the run-away condition. It may also be deduced from (2.86) and (2.88) that *the steady-state speeds are equal when:*

$$F = \frac{P_s A_{rod}}{2} \quad \text{where the rod area } A_{rod} = A_1 - A_2 \tag{2.89}$$

2.6 Undamped Natural Frequency of Actuators, the Effect of Fluid Compressibility and Load Mass

It is a simple matter to determine the undamped natural frequency of an actuator load since it is dominated by the interaction between fluid compressibility and load inertia in practice. Also many applications need to minimise the line volumes between actuator and servovalve and it is common to place the servovalve and manifold on top of the actuator as shown in Figure 2.33. Consider the actuator schematic in Figure 2.34 with the flows into and out of the actuator suddenly shut off, the connecting line volumes are negligible and the actuator volume changes are negligible under dynamic (oscillatory in this case) conditions.

We first consider the general flow continuity and force equations. For a general control volume V, the dynamic flow continuity equation representing the difference between the input flow rate Q_i and the output flow rate Q_o is given by

$$Q_i - Q_o = \frac{dV}{dt} + \frac{V}{\beta}\frac{dP}{dt} \tag{2.90}$$

Figure 2.33. High-performance servoactuator [Star Hydraulics Ltd UK, Eland Engineering Ltd UK]

The load in reality often has significant mass such that the force equation for a cylinder or the torque equation for a motor must be modified as follows:

$$\text{for a cylinder} \quad P_1 A_1 - P_2 A_2 = F + F_{\text{losses}} + M\frac{dU}{dt} \tag{2.91}$$

$$\text{for a motor} \quad D_m\left(P_1 - P_2\right) = T + T_{\text{losses}} + J\frac{d\omega}{dt} \tag{2.92}$$

where M is the load mass for a cylinder and J is the load inertia for a motor. Fluid bulk modulus β is discussed in more detail later in this chapter. Returning to the cylinder shown in Figure 2.34 and the application of (2.90) and (2.91) then gives:

$$0 = A_1 U + \frac{V_1}{\beta}\frac{dP_1}{dt}$$

$$0 = A_2 U - \frac{V_2}{\beta}\frac{dP_2}{dt} \tag{2.93}$$

$$P_1 A_1 - P_2 A_2 = M\frac{dU}{dt}$$

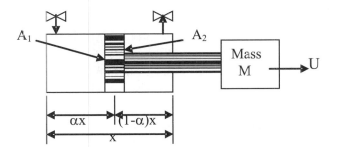

Figure 2.34. Sudden closure of an actuator

It is assumed that a load force has no effect on the frequency of oscillation for a linear system as in this case. It then follows that:

$$U + \frac{M}{\left[\dfrac{A_1^2\beta}{V_1} + \dfrac{A_2^2\beta}{V_2} \right]} \frac{d^2U}{dt^2} = 0 \qquad (2.94)$$

Hence the undamped natural frequency is given by:

$$\omega_n^2 = \frac{\beta A_2}{Mx} \left[\frac{\gamma}{\alpha} + \frac{1}{(1-\alpha)} \right] \qquad (2.95)$$

It will be deduced that *the natural frequency has a minimum value* as the position changes and this occurs when:

$$\alpha = \frac{\sqrt{\gamma}}{1+\sqrt{\gamma}} \quad \text{and with} \quad \frac{V_1}{V_2} = \gamma^{3/2} \qquad (2.96)$$

Typical actuator area ratios then leads to the following interpretation of this minimum natural frequency condition as shown in Table 2.1:

Table 2.1. Variation in position and volume ratio for minimum undamped natural frequency

γ	1.0	1.25	1.50	1.75	2.00
α	0.5	0.53	0.55	0.57	0.59
V_1/V_2	1.0	1.40	1.84	2.32	2.83

It can be seen from the data that for a double-rod actuator the minimum natural frequency occurs with the rod at its mid-position and with equal volumes either side. For the single-rod actuator the position does not vary significantly from the mid-position for a range of area ratios, although the volume ratio varies significantly with area ratio at the minimum frequency condition.

2.7 Some General Observations on Line Pressures in Servovalve/Actuator Control Systems in the Presence of Load Mass and Inertia

Some observations on dynamic behaviour can be made at this point, and for the particular, yet realistic, case of:

- a double rod actuator $A_1 = A_2 = A$
- equal initial volumes on either side $V_{10} = V_{20} = V_0$ for both a cylinder and a motor
- minor losses are neglected

For a cylinder drive

$$k_f i \sqrt{P_s - P_1} = AU + \frac{(V_0 + Ay)}{\beta} \frac{dP_1}{dt}$$

$$k_f i \sqrt{P_2} = AU - \frac{(V_0 - Ay)}{\beta} \frac{dP_2}{dt} \qquad (2.97)$$

$$(P_1 - P_2) A = F + M \frac{dU}{dt}$$

For a motor drive

$$k_f i \sqrt{P_s - P_1} = D_m \omega + \frac{V_0}{\beta} \frac{dP_1}{dt}$$

$$k_f i \sqrt{P_2} = D_m \omega - \frac{V_0}{\beta} \frac{dP_2}{dt} \qquad (2.98)$$

$$D_m (P_1 - P_2) = T + J \frac{d\omega}{dt}$$

It can be seen that by subtracting the two flow equations for both cases, then by induction, the sum of line pressures is dynamically constant and equal to supply pressure if the changes in volume during dynamic operation are negligible for a cylinder, as is often the case.

Dynamically $P_1 + P_2 \approx P_s$ \qquad (2.99)

This deduction is particularly appropriate to a motor drive where the lines either side will probably be of the same volume and, of course, the changes in volume during dynamic operation are negligible. Under dynamic conditions an increase in P_1 results in a similar decrease in P_2, a particular danger being cavitation under such conditions with large load forces and/or large moving masses.

2.8 Linearisation Technique to Estimate the Dynamic Behaviour of Nonlinear Systems

Now consider the servovalve and actuator simple circuit shown in Figure 2.35. This circuit will be used again when the influence of long lines is discussed.

Considering just the output port of the servovalve, then consideration of the linearisation technique discussed previously leads to the following equations:

$$Q = k_f i \sqrt{P_s - P} \tag{2.100}$$

$$\delta Q = k_i \delta i - k_p \delta P \tag{2.101}$$

$$k_i = \frac{\partial Q}{\partial i} = k_f \sqrt{P_s - P_{ss}} \quad k_p = -\frac{\partial Q}{\partial P} = \frac{Q_{ss}}{2(P_s - P_{ss})} \tag{2.102}$$

The actuator dynamics can be represented by the following linear differential equations with viscous friction, compressibility and load mass included:

$$\text{Force} \quad PA = B_v U + M \frac{dU}{dt} \tag{2.103}$$

$$\text{Flow} \quad Q = AU + \frac{V}{\beta} \frac{dP}{dt} \tag{2.104}$$

Linearisation of these equations leaves these forms unchanged:

$$\delta P A = B_v \delta U + M \frac{d \delta U}{dt} \tag{2.105}$$

$$\delta Q = A \delta U + \frac{V}{\beta} \frac{d \delta P}{dt} \tag{2.106}$$

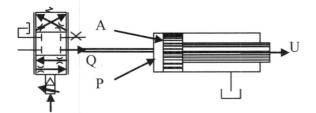

Figure 2.35. Servovalve and actuator coupled by a short line

Consequently all the system equations may be combined to produce in this example the differential equation relating δU to δi. This is given by:

$$\left(1+\frac{k_p B_v}{A^2}\right)\delta U+\left(\frac{VB_v}{\beta A^2}+\frac{k_p M}{A^2}\right)\frac{d\,\delta U}{dt}+\frac{MV}{A^2\beta}\frac{d^2\delta U}{dt^2}=\frac{k_i\delta i}{A} \tag{2.107}$$

Therefore it has been established that the velocity response to a change in servovalve current will be second-order and could be oscillatory depending upon the system parameters that exist. Usually for hydraulic systems in practice, damping is increased away from the steady-state condition when speeds are increased. Also in practice the δ notation is often omitted, though mathematically incorrect, for simplicity. Assume data are as follows:

- Line and actuator volume $V=1.42\times 10^{-3}\,\text{m}^3$ Actuator bore 100 mm
- Servovalve flow $=20\,\text{l/min}$, no load with a supply pressure of 70 bar, 15 mA
- Mineral oil density 860 kg/m^3 Total load mass $M=8500$ kg
- fluid effective bulk modulus $\beta=1.43\times 10^9\,\text{N/m}^2$
- Actuator viscous/coulomb friction damping coefficient $B_v=4.8\times 10^5\,\text{N/ms}^{-1}$

Evaluating the linearised differential equation at the anticipated maximum velocity of 0.035 m/s with a pressure of 21.5 bar results in the following linearised differential equation:

$$\left(1+\frac{k_p B_v}{A^2}\right)\delta U+\left(\frac{VB_v}{\beta A^2}+\frac{k_p M}{A^2}\right)\frac{d\delta U}{dt}+\frac{MV}{A^2\beta}\frac{d^2\delta U}{dt^2}=\frac{k_i\delta i}{A}$$
$$\downarrow \qquad\qquad \downarrow \quad\; \downarrow \qquad\qquad \downarrow \tag{2.108}$$
$$0.22 \qquad\; 0.0077 \;\; 0.0039 \qquad 137\times 10^{-6}$$

Collecting terms together and simplifying:

$$\delta U+0.01\frac{d\,\delta U}{dt}+1.12\times 10^{-4}\frac{d^2\delta U}{dt^2}=\frac{0.82U_{ss}}{i_{ss}}\delta i \tag{2.109}$$

Laplace transforming (2.109) and developing the transfer function, and neglecting initial conditions, results in:

$$\delta U(s)=\frac{\dfrac{0.82U_{ss}}{i_{ss}}\delta i(s)}{\left(1+0.01s+1.12\times 10^{-4}s^2\right)}=\frac{\dfrac{0.82U_{ss}}{i_{ss}}\delta i(s)}{\left(1+\dfrac{2\zeta s}{\omega_n}+\dfrac{s^2}{\omega_n^2}\right)} \tag{2.110}$$

It may be deduced that the undamped natural frequency $\omega_n=94.5$ rad/s (15 Hz) and a damping ratio $\zeta=0.47$ for small variations around maximum velocity. The exact solution and linearised solution are shown in Figure 2.36 for a change in demanded velocity from rest to maximum.

The linearised solution is remarkably close to the exact solution allowing for the differences in steady-state speeds as a result of linearisation. Comparisons, of course, are closer as the conditions of the linearisation process are better met such as lower steady-state speeds.

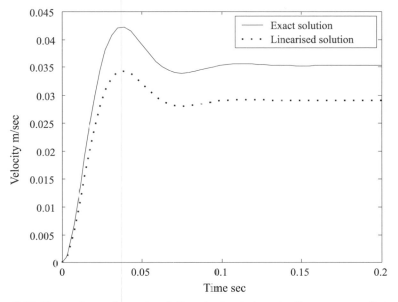

Figure 2.36. Comparison of exact and linearised solutions to the response of a servo-valve/cylinder drive

2.9 Underlapped Servovalve Spools – the Effect on Steady-state Behaviour for Closed-loop Control Systems

In closed-loop control systems, particularly position and pressure, the servovalve spool is ostensibly centred in the steady-state condition for the ideal case with no line leakages or transducer fault. It will be clear from earlier discussions that the linearised pressure coefficient is zero at this steady-state condition for a critically-lapped spool. Therefore the servovalve provides no damping in such conditions. Damping can be provided at this "null" condition by introducing spool underlap and it is not unusual for manufacturers to deliberately ensure this at the production stage if specified by the user. Therefore consider a position control system as shown in Figure 2.37.

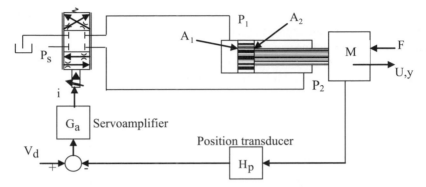

Figure 2.37. Servovalve and actuator position control system

For steady-state conditions $Q_1 = Q_2 = 0$, and from earlier work, a steady-state error in current can exist depending on the load force F. Considering the dominant load force:

$$P_1 A_1 - P_2 A_2 = F \qquad (2.111)$$

The steady-state condition for servovalve current, hence position error, is

$$\bar{i} = \frac{i}{i_u} = \frac{G_a H_p y_{error}}{i_u} = \frac{\sqrt{1+\overline{F}} - \sqrt{\gamma - \overline{F}}}{\sqrt{1+\overline{F}} + \sqrt{\gamma - \overline{F}}} \qquad \overline{F} = \frac{F}{P_s A_2} \qquad (2.112)$$

Therefore the position error can be driven to zero for one load condition only, which may be deduced from (2.112). At this optimum condition the two line pressures are equal, having values of half the supply pressure:

$$\overline{F} = \frac{(\gamma - 1)}{2}$$

$$F = \frac{P_s A_{rod}}{2} \quad \text{where the rod area } A_{rod} = A_1 - A_2 \qquad (2.113)$$

$$P_1 = P_2 = P_s / 2 \qquad (2.114)$$

In practice this optimum condition may be achieved by simply adjusting supply pressure until the line pressures are equal, providing this results in an acceptable operating condition. Consider measurements on such a system using the servovalve previously discussed and having an underlap-equivalent of ± 0.27 mA. Further details are:

- actuator bore diameter = 50.8 mm, rod diameter = 28.58 mm
- area ratio $\gamma = 1.463$
- servoamplifier gain G_a 10 mA/V
- position transducer gain H_p 0.0413 V/mm

At no load the position error is –0.062 mm. Since the actuator was connected to one arm of a manipulator, the actual end effector position error is significantly greater due to the moment arm effect. At optimum load for zero position error the load force F = 3200 N at a supply pressure of 100 bar and 1600 N at a supply pressure of 50 bar. Measured error current and actuator pressures are shown in Figure 2.38. These results show that the theory holds and the measured pressures are always slightly below the theoretical values. It can also be seen that the condition for zero position error agrees with that predicted. Closer observation of the steady-state position may well reveal small, apparently random, variations about the expected value. This may be attributed to, for example:

- servovalve electromagnetic stage hysteresis, typically < 3% of rated current
- inherent yet small drift of the spool
- system friction
- spool underlap not exactly matched at all ports

A common method of improving small steady-state fluctuations is to introduce a high frequency dither signal at the servoamplifier. In this case a frequency of

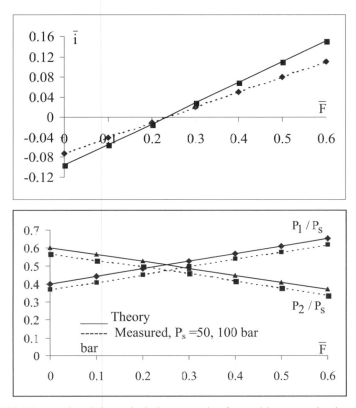

Figure 2.38. Measured and theoretical characteristics for position control using an underlapped servovalve

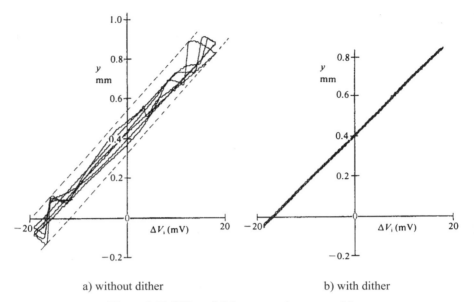

a) without dither   b) with dither

Figure 2.39. Effect of dither on steady-state position

150 Hz was used with an amplitude of 0.19 mA (recall that the underlap equivalent $i_u = 0.27$ mA). For this study, the result of applying a demanded change in position, at optimum load and around a value of 0.4 mm from the actuator central position, is shown in Figure 2.39 without and with dither.

The position change was created using a triangular shaped dither waveform, and three cycles are illustrated. It can be seen that without dither the variation in position is different for each cycle, and that the effect is virtually removed with the introduction of dither. The use of dither seems to be a matter of personal choice and also seems to be a matter of fashion.

2.10 Underlapped Servovalve Spools – the Effect on Dynamic Behaviour for Closed-loop Control Systems

First consider the linearised coefficients for the underlapped spool using the nonlinear flow equations (2.58) and (2.60). For closed-loop position control at optimum load such that there is no steady-state position error, the linearised coefficients become:

Flow gain $\qquad k_{i1} = k_{i2} = k_i = 2k_f\sqrt{\dfrac{P_s}{2}}$

Pressure coefficient $\quad k_{p1} = k_{p2} = k_p = \dfrac{k_f i_u}{\sqrt{\dfrac{P_s}{2}}}$ $\qquad$ (2.115)

The linearised transfer function for the servovalve/cylinder under position control may now be derived. A completely general solution leads to a complicated transfer function that is difficult to directly interpret so a general feel for closed-loop performance may be gained by making the following assumptions:

- **double-rod actuator ($\gamma=1$)** with **no load force ($F=0$)**, for example with the cylinder horizontal, hence zero position error
- **piston centralised,** the minimum undamped natural frequency condition
- the load has viscous damping B_v and moving mass M.

The system equations are:

$$Q_1 = k_f\left(i_u + i\right)\sqrt{P_s - P_1} - k_f\left(i_u - i\right)\sqrt{P_1} = AU + \frac{V}{\beta}\frac{dP_1}{dt}$$

$$Q_2 = k_f\left(i_u + i\right)\sqrt{P_2} - k_f\left(i_u - i\right)\sqrt{P_s - P_2} = AU - \frac{V}{\beta}\frac{dP_2}{dt}$$

(2.116)

$$\left(P_1 - P_2\right)A = B_v U + M\frac{dU}{dt}$$

Linearising these equations gives the following transfer function relating position, y, to input current, i:

$$\delta y(s) = \frac{\dfrac{k_i \delta i(s)}{A}}{s\left(1 + \dfrac{R_v}{2R_u} + s\left(\dfrac{L}{2R_u} + \dfrac{CR_v}{2}\right) + s^2\dfrac{LC}{2}\right)} = G_{act}(s)\,\delta i(s)$$

(2.117)

$$R_u = 1/k_p \quad L = M/A^2 \quad R_v = B_v/A^2 \quad C = A\ell/\beta \quad \ell = \text{half stroke}$$

It can be seen that both viscous damping and spool underlap contribute towards the total damping. The closed-loop block diagram may now be constructed as shown as Figure 2.40.

To determine stability we consider the system characteristic equation:

$$1 + G(s)H(s) = 0$$

(2.118)

$$1 + G_a\,G_{act}(s)\,H_p = 0$$

$$s^3\frac{LC}{2} + s^2\left(\frac{L}{2R_u} + \frac{CR_v}{2}\right) + s\left(1 + \frac{R_v}{2R_u}\right) + \frac{G_a H_p k_i}{A} = 0$$

(2.119)

$$a_3 s^3 + a_2 s^2 + a_1 s + a_0 = 0$$

Stability may be studied using the Routh Array method which results in the following condition for this third-order system to be stable:

$$a_2 a_1 > a_3 a_0$$

$$\left(\frac{L}{2R_u} + \frac{CR_v}{2}\right)\left(1 + \frac{R_v}{2R_u}\right) > \frac{LC}{2}\frac{G_a H_p k_i}{A}$$

(2.120)

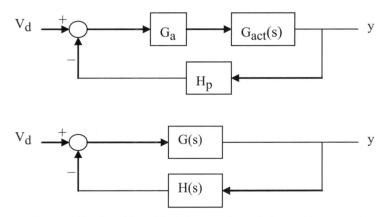

Figure 2.40. Closed-loop block diagram for cylinder position control

If it also assumed that the viscous damping term is negligible then $R_v = 0$ and from (2.120):

$$i_u > G_a H_p \ell \frac{P_s}{\beta} \qquad (2.121)$$

Some comments are appropriate here:

- This simple expression for the required underlap is readily achieved by the appropriate selection of system parameters. It will be notices that $G_a H_p \ell$ is the current generated by the electronics for 1/2 stroke and it is perhaps a good design feature to make this value equal to the servovalve rated current.
- Stability under closed-loop control is independent of load mass and the minimum value of bulk modulus appropriate to operation must be used.
- Underlap is therefore a good, and guaranteed, way to add damping to a servovalve controlled actuator although it must be realised that this is at an additional energy cost due to the extra leakage created by the underlap.
- From (2.121) it would seem that only a very small value of underlap is necessary for stability due to the typical value of P_s/β. In fact, the underlap value may be selected to give a desired closed-loop transient response.

The closed-loop transfer function is given by

$$\frac{\delta y(s)}{\delta y_d(s)} = \frac{\dfrac{G_a H_p k_i}{A}}{\dfrac{G_a H_p k_i}{A} + s\left(1 + \dfrac{R_v}{2R_u}\right) + s^2\left(\dfrac{L}{2R_u} + \dfrac{CR_v}{2}\right)} \qquad (2.122)$$

where $\delta y_d(s)$ is the demandes position.

This transfer function may be expressed in a standard form using an appropriate closed-loop transient response criterion such as the:

"*Integral of Time multiplied by Absolute Error*" (ITAE) criterion

For a third-order transfer function this ITAE form is given by:

$$\frac{\delta y(s)}{\delta y_d(s)} = \frac{\omega_o^3}{\omega_o^3 + 2.15\omega_o^2 s + 1.75\omega_o s^2 + s^3} \tag{2.123}$$

Assuming that viscous damping is negligible, for this underlap resistance-dominated problem, then gives the required system parameters:

$$K = \frac{G_a H_p k_i}{A} \qquad \omega_o = 2.15K \qquad \frac{2}{LC} = 9.94K^2 \qquad \frac{k_p}{C} = 3.76K \tag{2.124}$$

Whether or not these constraints can be achieved depends upon the components chosen. Should the design be possible then the transient response to a step input would have the third-order ITAE shape shown in Figure 2.41.

The time response shown in Figure 2.41, plotted against a non-dimensional time, has the desired overshoot but in practice may not be fast enough for an electrohydraulic control system, particularly for small values of the gain K. This system design approach using spool servovalve underlap could therefore be a disadvantage in some applications. If underlap is not present, then clearly this

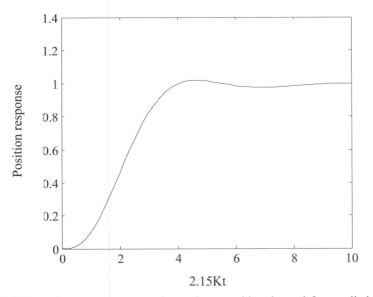

Figure 2.41. Transient response to a unit step input position demand for a cylinder drive designed to match the third-order ITAE criterion

closed-loop position system requires viscous damping to sure stability. From (2.80) this condition is:

$$B_v > G_a H_p \frac{M k_f \sqrt{2P_s}}{A} \tag{2.125}$$

Stability now depends upon load mass, the servovalve flow coefficient, and the actuator area, all new terms that were not present when underlap was assumed dominant.

2.11 Proportional Pressure Relief Valve Modelling Concepts

Pressure relief valves can be complex components requiring detailed consideration of their operating dynamic characteristics, particularly if they are operated via proportional solenoid technology. The dynamic control voltage/pressure characteristic is required and the type shown in Figure 2.42 serves to illustrate various practical issues for modelling a two-stage pressure relief valve.

Defining the performance of this valve now becomes more complex due to the electromagnetic stage whose dynamics may not be neglected in comparison to the drive stage of the servovalve previously discussed. It becomes obvious that much experimental testing of the valve is needed to obtain sufficiently accurate data to allow the resulting model to be used in a system simulation. Some of the valve equations are well-established and now follow:

$$\text{Flow continuity} \quad Q_p = Q_{rv} + A\frac{dy}{dt} + \frac{V_s}{\beta}\frac{dP_s}{dt} \tag{2.126}$$

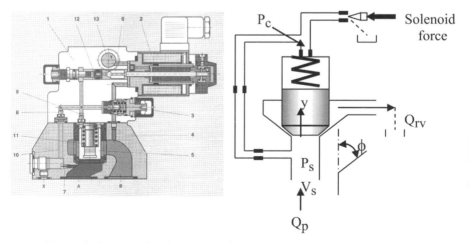

Figure 2.42. Proportional pressure relief valve [Courtesy Bosch Rexroth Ltd]

Force balance $\left(P_s - P_c\right)A = F_o + ky + \alpha a_c P_s + B\dfrac{dy}{dt} + m\dfrac{d^2y}{dt^2}$ (2.127)

$Q_{rv} = C_q a_c \sqrt{\dfrac{2P_s}{\rho}}$ Port area $a_c = \pi c_s y \sin\varphi$ (2.128)

- F_o = initial spring force
- d_s = poppet diameter
- B = poppet viscous damping coefficient

k = spring stiffness
A = poppet area
m = poppet mass

The flow reaction force term, $\alpha a_c P_s$, contains the constant α which must be experimentally determined since its value depends upon the flow port geometry and the flow regime in place over the flow range of the application.

The general form of the flow reaction force is based upon well-established momentum change theory but its precise magnitude is always prone to variations in practice. In this study the determination of α is easily achieved since the back pressure P_c can be measured during steady-state conditions since it follows that:

$$\dfrac{d\left(P_s - P_c\right)}{dQ_{rv}} = \dfrac{P_s}{C_q A\sqrt{\dfrac{2P_s}{\rho}}}\left[\dfrac{k}{P_s \pi d_s \sin\phi} + \alpha\right]$$ (2.129)

Table 2.2. Poppet flow reaction force coefficient evaluation

Supply pressure P_s (bar)	30	40	50	60
Coefficient α	2.24	2.36	2.73	2.49

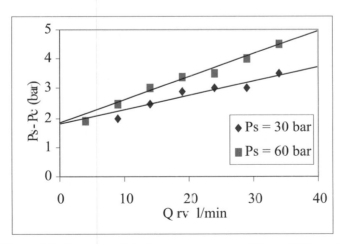

Figure 2.43. Pressure relief valve poppet pressure differential/flow rate

Equation (2.129) is evaluated from the graph of $(P_s - P_c)/Q_{rv}$ and the value of α determined from the slope at constant supply pressure. The graph is shown as Figure 2.43 and Table 2.2 shows the results.

These values are reasonably consistent, given the small pressure difference measurement accuracy, and indicate flow reaction forces typically twice that predicted from ideal momentum change theory.

The solenoid force characteristic must also be obtained experimentally and this may be done by coupling the unit to a load cell and measuring the transient behaviour which is shown in Figure 2.44.

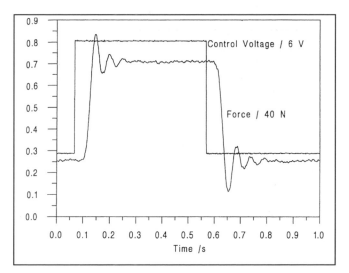

Figure 2.44. Solenoid measured force characteristic

From Figure 2.44 it is observed that a pure delay of 0.02 seconds exists due to electromagnetic effects. In addition it is found that the transient part may be ade-quately represented by a second-order dynamic characteristic having a damping ratio $\zeta = 0.3$ and an undamped natural frequency $\omega_n = 20\,\text{Hz}$. Hence a linear trans-fer function is used to represent this characteristic. It is assumed that the pilot pressure P_c is varied via a voltage applied to the solenoid stage:

$$V(s) = \frac{e^{-sT}}{\left(1 + \dfrac{2\zeta}{\omega_n}s + \dfrac{s^2}{\omega_n^2}\right)} V_{ref}(s) \tag{2.130}$$

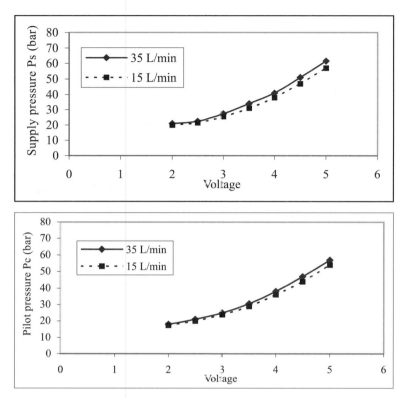

Figure 2.45. Steady-state variation of pilot pressure and supply pressure with control voltage

where T is the pure delay of 0.02 seconds. The steady-state pressure characteristic was found experimentally to be of the following form:

$$P_c \text{ or } P_s = b_o + b_1 V + b_2 V^2 \tag{2.131}$$

The experimental evidence for this is shown in Figure 2.45 for a range of valve flow rates.

Bringing together the various equations and nonlinear terms then allows the open-loop performance to be evaluated once a suitable numerical solution technique has been selected. A comparison between measured and modelled transient pressures is shown in Figure 2.46 for demanded supply pressure change from 27 bar to 65 bar.

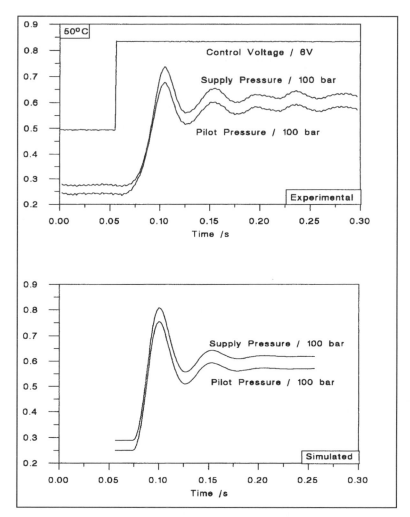

Figure 2.46. Comparison of measured and simulated pressure responses for a proportional pressure relief valve

2.12 Long Lines

2.12.1 The Basic Equations

In many applications the pump and associated control valve block may be separated from the actuator by a distance of many metres. For example in forging press and steel mill applications the working environment is not conducive to electrohydraulic components or the extremely large power pack may simply be unacceptable at the working site. In these applications the line cannot be considered as

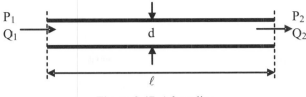

Figure 2.47. A long line

a lumped volume as part of a flow continuity equation such as equation (2.1). New equations must be incorporated to describe the complex wave propagation dynamics and a great deal of research has been undertaken in this area. Considering Figure 2.47.

The velocity of sound C_o for a wave propagating through a fluid is given by:

$$C_o = \sqrt{\frac{\beta}{\rho}} \qquad (2.132)$$

where for a typical mineral oil within a rigid steel line C_o is typically 1300 m/s. The time for a wave to travel down the line, often termed the delay time T, and the blocked line fundamental resonant frequency, f, are:

$$T = \frac{\ell}{C_o} \qquad f = \frac{C_o}{2\ell} \qquad (2.133)$$

This frequency can be compared with the expected system natural frequency to decided whether or not a transmission line model is required. Solution of the basic transmission line equations leads to:

$$P_2 = \frac{\left(P_1 + Z_c Q_1\right)e^{-\Gamma\ell}}{2} + \frac{\left(P_1 - Z_c Q_1\right)e^{+\Gamma\ell}}{2} \qquad (2.134)$$

$$Q_2 = \frac{\left(Q_1 + P_1/Z_c\right)e^{-\Gamma\ell}}{2} + \frac{\left(Q_1 - P_1/Z_c\right)e^{+\Gamma\ell}}{2} \qquad (2.135)$$

- Z_c = Characteristic impedance
- Γ = Propagation constant

The characteristic impedance and propagation constant are complex functions depending upon the line model chosen. Using the following general notation for the "per unit length" line parameters for resistance, inductance, capacitance:

$$R' = \frac{128\mu}{\pi d^4} \qquad L' = \frac{\rho}{a} \qquad C' = \frac{a}{\beta} \qquad (2.136)$$

Series impedance $Z = (R' + sL')$ $\qquad (2.137)$

Shunt admittance $Y = sC'$ $\qquad (2.138)$

Characteristic impedance $Z_c = \sqrt{\dfrac{Z}{Y}}$ (2.139)

Propagation constant $\Gamma = \sqrt{ZY}$ (2.140)

- **lossless model** $R' = 0$

  Characteristic impedance $Z_c = Z_{ca} = \sqrt{\dfrac{\rho\beta}{a^2}}$ (2.141)

  Propagation constant $\Gamma = \Gamma_a = \dfrac{s}{C_o}$ (2.142)

- **average friction model**

  Characteristic impedance $Z_c = \sqrt{\dfrac{R' + sL'}{sC'}}$ (2.143)

  Propagation constant $\Gamma = \sqrt{(R' + sL')sC'}$ (2.144)

- **distributed friction model**

  Series impedance $Z = \dfrac{sL'}{1 - \dfrac{2J_1(f)}{f\,J_o(f)}}$ $f = j\sqrt{\dfrac{sr^2}{\nu}}$ (2.145)

  Shunt admittance $Y = sC'$ (2.146)

2.12.2 Application to a Servovalve/Single-line/Cylinder

For the purpose of example, consider a servovalve with one line coupled to an actuator as shown in Figure 2.48. The lumped volume solution together with a linearised solution was analysed in Section 2.8.

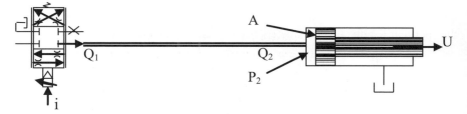

Figure 2.48. Transmission line connecting the flow source to an actuator

Data used are as follows:

- servovalve flow rate of 20 L/min with a pressure differential of 70 bar
- supply pressure 70 bar
- line length $\ell = 10.73$ m, diameter $= 13$ mm
- mineral oil density 860 kg/m^3
- kinematic viscosity $v = 38$ cSt
- fluid effective bulk modulus $\beta = 1.43 \times 10^9$ N/m^2
- actuator bore 100 mm
- total load mass M $= 8500$ kg
- actuator effective velocity/coulomb damping coefficient B $= 4.8 \times 10^5$ N/ms^{-1}

Before the solution to the transmission line equations is pursued it is useful to *consider a more reasonable lumped analysis* beyond simply fluid compressibility effects as discussed earlier. In this case the line is represented by a better lumped approximation to its properties of resistance R, inductance L and capacitance C. For example, these line properties given by (2.136) when expressed for the entire line become:

$$R = \frac{128\mu\ell}{\pi d^4} \qquad L = \frac{\rho\ell}{a} \qquad C = \frac{a\ell}{\beta} \qquad (2.147)$$

$$R = 5.05\times10^8 \; \text{Nm}^{-2}/\text{m}^3\text{s}^{-1}, \; L = 6.94\times10^7 \; \text{kg}\,\text{m}^{-4}, \; C = 10^{-12} \; \text{m}^3/\text{N}\,\text{m}^{-2}$$

One approach is to distribute these resistance, inductance and capacitance properties along the line in a symmetrical "lumped" way such as the use of two symmetrical lumps (sometimes referred to as π networks). Each π network has capacitance place either side of a series resistance and inductance as shown in Figure 2.49.

This results in three flow continuity equations and two force equations for the line, and which may easily included in a computer simulation. The use of more lumps, say 4, 6, 8, etc.) may not necessarily improve the modelling accuracy and some care must be taken, particularly when arranging the force continuity equations. However, it is to be expected that the model described will be better than

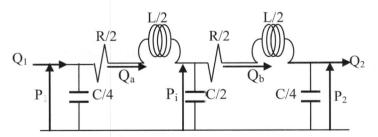

Figure 2.49. A line represented as two lumped symmetrical networks

a single lump up to the first modal frequency. The system mathematical model then becomes:

Servovalve $Q_1 = k_f i \sqrt{P_s - P_1}$ $\quad\quad$ (2.148)

Line equations

$$Q_1 - Q_a = \frac{C}{4} \frac{dP_1}{dt}$$

$$P_1 - P_i = \frac{R}{2} Q_a + \frac{L}{2} \frac{dQ_a}{dt}$$

$$Q_a - Q_b = \frac{C}{2} \frac{dP_i}{dt} \quad\quad (2.149)$$

$$P_i - P_2 = \frac{R}{2} Q_b + \frac{L}{2} \frac{dQ_b}{dt}$$

$$Q_b - Q_2 = \frac{C}{2} \frac{dP_2}{dt}$$

Load flow $Q_2 = AU + \dfrac{V}{\beta} \dfrac{dP_2}{dt}$ $\quad\quad$ (2.150)

Load force $P_2 A = BU + M \dfrac{dU}{dt}$ $\quad\quad$ (2.151)

Returning now to the transmission equations, the simplest pair of equations evolves from *the assumption of a lossless line*, for example large diameter steel pipes where pressure drops due to friction are designed to be a minimum, preferably negligible. The design study under discussion here, for example, has only a 1.4 bar line pressure drop using the data to follow later. It can be shown that the lossless line equations may be written in the following form suitable for computer simulation:

$$P_2 + Z_{ca} Q_2 = z^{-1} \left(P_1 + Z_{ca} Q_1 \right) \quad\quad (2.152)$$

$$P_1 - Z_{ca} Q_1 = z^{-1} \left(P_2 - Z_{ca} Q_2 \right) \quad\quad (2.153)$$

Z_{ca} = lossless line characteristic impedance

$$\sqrt{\frac{\rho \beta}{a^2}} = 0.83 \times 10^{10} \ \mathrm{Nm^{-2} m^3 s^{-1}} \quad\quad (2.154)$$

$z^{-1} = e^{-sT}$ represents the line delay of T seconds and a is the line cross-sectional area. In other words, the left-hand sides are evaluated from the right-hand side one delay interval later.

To consider *the distributed friction model* we will use the *modal analysis* technique which uses transfer function approximations to the exact solutions of the transmission line functions in the frequency domain. This approach provides a computationally stable solution that is easily implemented within a hydraulic circuit. Equations (2.134) and (2.135) are written in the hyperbolic function form as follows:

$$\begin{bmatrix} P_2 \\ Q_2 \end{bmatrix} = \begin{bmatrix} \cosh \Gamma \ell & -Z_c \sinh \Gamma \ell \\ -\sinh \Gamma \ell / Z_c & \cosh \Gamma \ell \end{bmatrix} \begin{bmatrix} P_1 \\ Q_1 \end{bmatrix} \tag{2.155}$$

For hydraulic circuit simulation it is convenient to use the following pair of equations derived from (2.155):

$$P_1 = \frac{Z_c \cosh \Gamma \ell}{\sinh \Gamma \ell} \left[Q_1 - \frac{Q_2}{\cosh \Gamma \ell} \right] \tag{2.156}$$

$$P_2 = \frac{Z_c \cosh \Gamma \ell}{\sinh \Gamma \ell} \left[\frac{Q_1}{\cosh \Gamma \ell} - Q_2 \right] \tag{2.157}$$

Using the modal approximations for the hyperbolic functions then allows the pair of equations to be approximated by a suitable combination of second-order transfer functions using the following structure:

$$P_1(s) = \frac{T_2(s)}{sC} \left[Q_1 - T_1(s) Q_2 \right] \tag{2.158}$$

$$P_2(s) = \frac{T_2(s)}{sC} \left[T_1(s) Q_1 - Q_2 \right] \tag{2.159}$$

The modal constants may be obtained from the original work by Hsue and Hullender (1983). In practice an even number of modes are selected and choosing $n=4$ or 6 modes gives good results, often showing little improvement with the addition of higher mode pairs. Note that the $T_1(s)$ transfer function must be normalised such that the steady-state transmission gain is unity. The modal functions are given by:

$$T_1(s) = \frac{1}{\cosh \Gamma \ell} \approx \sum_{i=1}^{n\,\text{modes}} \frac{a_i \bar{s} + b_i}{\bar{s}^2 + 2\zeta_i \omega_{ni} \bar{s} - \omega_{ni}^2} \tag{2.160}$$

$$T_2(s) = \frac{Z_c \cosh \Gamma \ell}{\sinh \Gamma \ell} \approx 1 + \sum_{i=1}^{n\,\text{modes}} \frac{\dfrac{D_n a_i}{Z_{ca}} \bar{s}^2 + \dfrac{D_n b_i}{Z_{ca}} \bar{s}}{\bar{s}^2 + 2\zeta_i \omega_{ni} \bar{s} + \omega_{ni}^2} \tag{2.161}$$

$$Z_{ca} = \sqrt{\frac{\rho \beta}{a^2}} = 0.83 \times 10^{10} \quad C = \frac{a\ell}{\beta} = 10^{-12} \quad C_o = \sqrt{\frac{\beta}{\rho}} = 1290\,\text{m/s} \tag{2.162}$$

$$D_n = \frac{\nu \ell}{C_o r^2} = 0.00756 \quad \bar{s} = s\tau \quad \tau = \frac{r^2}{\nu} = 1.1$$

Each set of modal constants is evaluated using the root index λ_c:

for $T_1(s)\lambda_c = (i-1/2)/D_n$

for $T_2(s)\lambda_c = I/D_n$ (2.163)

where i is the mode number, in this study i = 1,2,3,4.

The $T_1(s)$ and $T_2(s)$ transfer functions for the example circuit are shown in Figure 2.50.

Returning to the problem, it is possible to compare the four solutions. The steady-state velocity U = 0.035 m/s and the steady-state pressure is 21.5 bar just to overcome the viscous friction and Coulomb friction force. The pure delay T = 8.32 ms and therefore the line fundamental resonant frequency is 60 Hz which is not too remote from the estimated lumped natural frequency of 13.6 Hz. Figure 2.51 shows a comparison between the various models. It can be seen that the lumped approximation has actually produced a result close to the distributed friction solution. The initial pure delay of 8.32 ms is of course better represented by either the lossless line model or the distributed friction model.

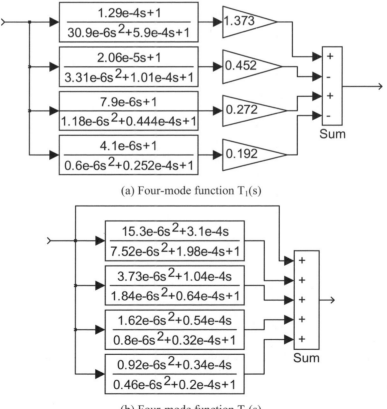

(a) Four-mode function $T_1(s)$

(b) Four-mode function $T_2(s)$

Figure 2.50. Modal transfer functions assuming n = 4 modes for the example circuit

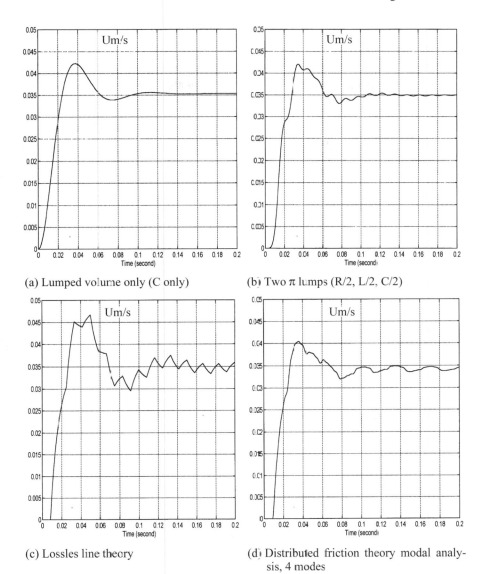

(a) Lumped volume only (C only)

(b) Two π lumps (R/2, L/2, C/2)

(c) Lossles line theory

(d) Distributed friction theory modal analysis, 4 modes

Figure 2.51. Comparison of line models for the servovalve/actuator circuit

The lossless line model is perhaps not to be recommended, particularly for large diameter lines having a small dissipation number D_n. Low values of D_n may also create simulation problems when using the modal analysis approach since a large number of modes will probably be required. In this case, interestingly, a two-lumped approach can produce acceptable results.

2.12.3 Modelling Actuator Volume Effects

Actuator volume effects may easily be taken into account using the previously described modal analysis approach. Consider the control system of Figure 2.52.

Combining the modal equations (2.160) and (2.161) with the actuator flow continuity equation for the extending case gives:

$$Q_a = A_1 U + \frac{(V_{10} + A_1 y)}{V_{line}} T_2(s) \left[T_1(s) Q_1 - Q_a \right] \tag{2.164}$$

$$P_a = \frac{1}{sC} T_2(s) \left[T_1(s) Q_1 - Q_a \right] \tag{2.165}$$

$$P_1 = \frac{1}{sC} T_2(s) \left[Q_1 - T_1(s) Q_a \right] \tag{2.166}$$

$$Q_1 = k_f i \sqrt{P_s - P_1} \tag{2.167}$$

$$Q_b = A_2 U - \frac{(V_{20} - A_2 y)}{V_{line}} T_2(s) \left[Q_b - T_1(s) Q_2 \right] \tag{2.168}$$

$$P_b = \frac{1}{sC} T_2(s) \left[Q_b - T_1(s) Q_2 \right] \tag{2.169}$$

$$P_2 = \frac{1}{sC} T_2(s) \left[T_1(s) Q_b - Q_2 \right] \tag{2.170}$$

$$Q_2 = k_f i \sqrt{P_2} \tag{2.171}$$

$$P_a A_1 - P_b A_2 = BU + M \frac{dU}{dt} \tag{2.172}$$

The initial actuator volumes are V_{10} and V_{20} and the line volume is V_{line}. These equations may be directly linked and, interestingly, require no volume inversion as used in conventional modelling of fluid compressibility.

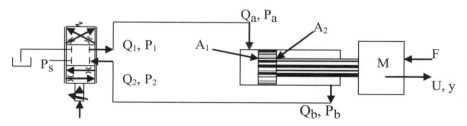

Figure 2.52. Servovalve and actuator coupled by long lines

2.12.4 Frequency Response

Frequency response techniques can be useful for analysing transmission line problems and can give some valuable insights into the stability of closed-loop systems. For example, consider testing a line by simply coupling it to a servovalve as shown in Figure 2.53.

A frequency response test is easily achieved at different mean flow rates by superimposing a small-amplitude sinusoidal signal. Alternatively, a pseudo-random binary signal (PRBS) may be applied to speed-up the experimental process. In fact the latter method has been found preferable for a range of frequency response applications in fluid power. Frequency response methods historically preceded time domain methods because of the ability to express the hyperbolic form of the line equations into the frequency domain relatively easily. For the configuration shown in Figure 2.53, and recalling the servovalve linearised coefficients, the linearised transfer function becomes:

$$\frac{\delta\left(P_1 - P_2\right)}{2R_v k_i \delta i} = \frac{\dfrac{Z_c}{R_v}\left[\cosh \Gamma \ell + \dfrac{Z_c}{R_v}\sinh \Gamma \ell - 1\right]}{\left[\dfrac{2Z_c}{R_v}\cosh \Gamma \ell + \left(1 + \dfrac{Z_c^2}{R_v^2}\right)\sinh \Gamma \ell\right]} \qquad (2.173)$$

Details of this test are:

- line is 21.46 m long, 7 mm internal diameter, fluid temperature 45°C
- bulk modulus $\beta = 1.33 \times 10^9 \, \text{N/m}^2$, density 860 kg/m^3
- measured line resistance $R = 1.08 \times 10^{10} \, \text{Nm}^{-2}/\text{m}^3\text{s}^{-1}$.

Results are shown in Figure 2.54 for two conditions having a mean servovalve current of 1.5 mA and 6 mA. For these two conditions the impedance ratios are:

- @ 1.5 mA $Z_{ca}/R_v = 0.128$ $R/Z_{ca} = 0.388$
- @   6 mA $Z_{ca}/R_v = 0.441$ $R/Z_{ca} = 0.388$

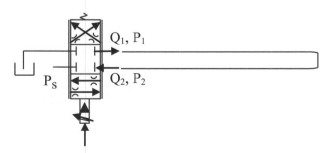

Figure 2.53. Servovalve coupled by a long line

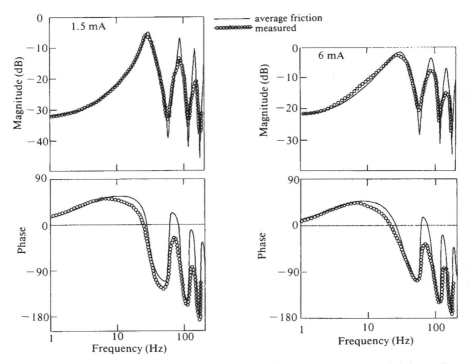

Figure 2.54. Frequency response of a sevovalve with output ports coupled by a line 21.46 m long, 7 mm internal diameter

Over the frequency range of interest the servovalve dynamics must be included and for this case the manufacturer's recommended second-order transfer function was included as follows:

$$G_{servovalve} = \frac{1}{1 + \dfrac{2\zeta s}{\omega_n} + \dfrac{s^2}{\omega_n^2}} \quad \zeta = 0.9 \quad \omega_n = 140\,\text{Hz} \qquad (2.174)$$

The average friction theory compare well with measurements, particularly up to the first mode where servovalve dynamic effects are small.

2.12.5 Frequency Response with Actuator Included

An actuator is now added to the servovalve/line to form a conventional closed-loop drive as shown in Figure 2.55.

Considering a linearised analysis, it will be recalled that the servovalve equations are written for null conditions as follows:

$$\delta Q_1 = k_i \delta i - k_p \delta P_1$$
$$\delta Q_2 = k_i \delta i + k_p \delta P_2 \qquad (2.175)$$

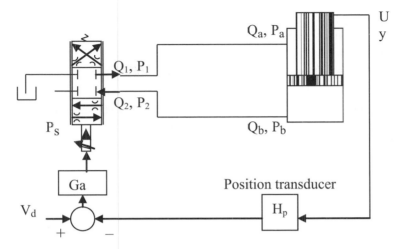

Figure 2.55. Servovalve/actuator coupled by long lines

Using the line hyperbolic equations (2.155), it is an easy matter to show that the servovalve linearised flow equations may be reflected to the actuator-end using the following transformations:

$$\delta Q_a = \frac{1}{T_1} k_i \delta i - \frac{T_2}{T_1} k_p \delta P_a$$

$$\delta Q_b = \frac{1}{T_1} k_i \delta i + \frac{T_2}{T_1} k_p \delta P_b$$

(2.176)

$$T_1 = \cosh \Gamma \ell + \frac{Z_c}{R_v} \sinh \Gamma \ell$$

$$T_2 = \cosh \Gamma \ell + \frac{R_v}{Z_c} \sinh \Gamma \ell$$

(2.177)

Applying these equations to the general case of a single-rod actuator is not strictly correct, apart from one unique load force, due to the varying flow gains for extension and retraction. However for the purpose of testing consider:

- A horizontal actuator with no load force and a small load mass.
- The flow gains are given by (2.84) and (2.86) for the extending and retracting cases. They are different due to the $\sqrt{\gamma}$ increase for the extending case. An average gain is assumed.
- For the frequency response test, in this case with PRBS excitation, a mean flow gain may be assumed.
- The frequency response is obtained by placing the actuator in closed-loop position control.

- The pressure coefficients are zero for a critically-lapped servovalve at the zero mean-current condition under position control.
- The actuator is at its mid-stroke position.
- The actuator bore = 50.8 mm diameter, rod diameter = 28.6 mm.
- Area ratio $\gamma = 1.463$.
- Each line is 10.73 m long, 7 mm internal diameter, fluid temperature 45°C.

The simplified open-loop transfer function, evaluated from closed-loop position control tests, is given by:

$$G(s) = \frac{K}{sT_1} \qquad T_1 = \cosh \Gamma \ell + \frac{Z_c}{R_v} \sinh \Gamma \ell \qquad (2.178)$$

The gain K is the assumed average including the position transducer gain of 41.3 V/m. Figure 2.56 shows the test result and Figure 2.57 shows the smoothed data with phase rearranged to correspond with the theoretical prediction.

The servovalve and lines are the same as those used in the previous example, the actuator actually being placed at the middle of the original 21.46 m long line. The theory given by (2.178) is therefore modified to also include the servovalve frequency response given by (2.174). Comparisons between practice and theory are good and the line resonant frequencies have clearly been detected and predicted.

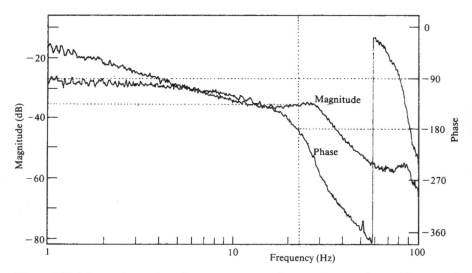

Figure 2.56. Measured open-loop frequency response for an actuator under position control and including long lines

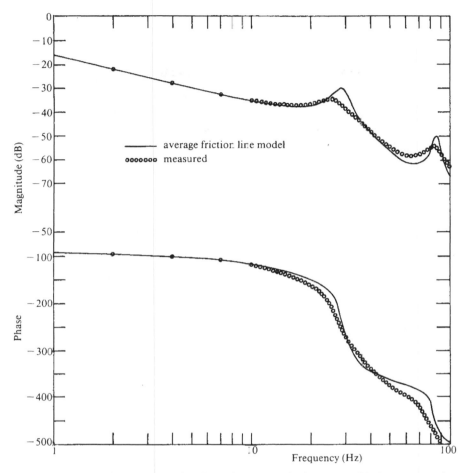

Figure 2.57. Open-loop transfer function of a servovalve/actuator with long connecting lines under position control

The ability to use such a system model is particularly important when closed-loop predictions need to be made. For example for this study Table 2.3 shows a comparison between different assumptions for predicting the servoamplifier gain and resulting frequency just at the point of closed-loop instability with position feedback using a linear variable differential transformer transducer. This was done in theory and practice by applying small changes in demanded position around the actuator central position. Position, velocity and pressure transducer responses must be carefully observed during these tests to just detect the onset of instability.

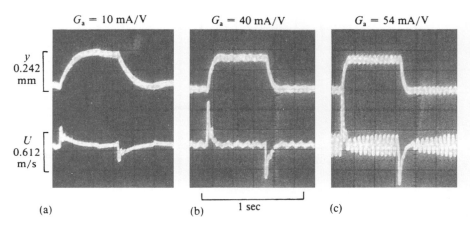

Figure 2.58. Position and velocity step responses for a cylinder position control system with long lines

Table 2.3. Conditions at the point of closed-loop instability

Approach	Servoamplifier gain (mA/V)	Frequency (Hz)
Measured	40.0	23.8
Average friction theory	43.0	25.2
Lossless line theory	12.3	28.7
Lines-neglected theory	532	47.0

The small-signal transient responses for position and velocity are shown in Figure 2.58 for an increasing servoamplifier gain. Permanent oscillations begin for a gain of 40 mA/V and are best observed from either the velocity recordings or the pressure recordings, not shown here. A change in audible noise at the onset of instability may also be noticed.

2.12.6 Response and Stability of a Servovalve/Motor Speed Control System with Long Lines

Now consider the motor speed control system shown in Figure 2.59.

Following established line and motor theory previously considered, the system equations are:

For the lines:

$$\delta Q_1 = k_i \delta i - k_p \delta P_1$$
$$\delta Q_2 = k_i \delta i + k_p \delta P_2$$

(2.179)

$$\delta Q_a = \frac{1}{T_1} k_i \delta i - \frac{T_2}{T_1} k_p \delta P_a$$

$$\delta Q_b = \frac{1}{T_1} k_i \delta i + \frac{T_2}{T_1} k_p \delta P_b \tag{2.180}$$

$$T_1 = \cosh \Gamma \ell + \frac{Z_c}{R_v} \sinh \Gamma \ell$$

$$T_2 = \cosh \Gamma \ell + \frac{R_v}{Z_c} \sinh \Gamma \ell \tag{2.181}$$

For the motor:

$$\delta Q_a = D_m \delta \omega + \frac{\delta (P_a - P_b)}{R_i} + \frac{\delta P_a}{R_e} \tag{2.182}$$

$$\delta Q_b = D_m \delta \omega + \frac{\delta (P_a - P_b)}{R_i} - \frac{\delta P_b}{R_e} \tag{2.183}$$

$$D_m \delta (P_a - P_b) = T + T_{cf} + B_v \omega + J \frac{d \, \delta \omega}{dt} \tag{2.184}$$

Here for the unloaded motor, $T = 0$ and the viscous damping term $B_v \omega$ is considered negligible in comparison to the servovalve damping contribution under speed control. The nonlinear torque friction characteristic of the motor, T_{cf}, has been measured and is shown in Figure 2.60.

In Figure 2.60, the small viscous friction torque effect has been removed. This was found to be equivalent to 1.2 Nm @60 rad/s giving a viscous coefficient of $B_v = 0.02$ Nm/rad s^{-1}. The motor displacement $D_m = 2.61 \times 10^{-6}$ m^{3}s^{-1}/rad and the dominant motor leakage resistance $R_e = 3 \times 10^{12}$ Nm^{-2}/m^{3}s^{-1}. It can also be seen

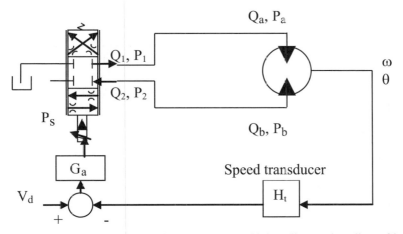

Figure 2.59. Servovalve/motor speed control system with long lines and nonlinear friction

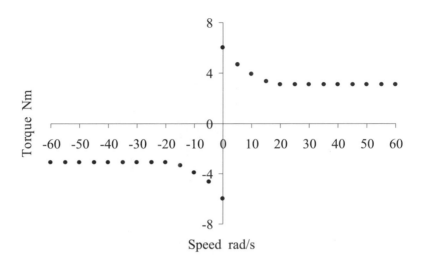

Figure 2.60. Measured motor friction torque for no load

that stiction torque is 6 Nm and the coulomb friction torque is 3 Nm. The system equations were solved using the method of characteristics (Watton, 1986) and with an exponential approximation for the stiction→Coulomb friction nonlinear characteristic.

Consider first the open-loop system. Some measured and simulated results for two different diameter lines are shown in Figure 2.61.

For these comparisons the servovalve current was suddenly lowered from 6 mA to 1.5 mA, representing a speed change from 51 rad/s to 13 rad/s. It can be seen that the 13 mm diameter lines result in stick-slip oscillations, which do not exist for the 7 mm diameter lines. It is confirmed that coulomb friction is a stabilising influence but stiction is a destabilising influence. For this steady-state condition, the simulation may be used to construct stick-slip and cavitation boundaries for a range of line lengths and diameters, and the results are shown in Figure 2.62.

It has been shown in simulation studies that stick-slip oscillation will not exist for line diameters below 8 mm, whatever the length. Selecting the 7 mm diameter lines case, the closed-loop system behaviour may be studied. Figure 2.63 shows the closed-loop block diagram using the previously derived linearised equations.

The friction nonlinearity has been put into the specific form shown so that the describing function (DF) method may be used to analyse the condition for closed-loop stability. The particular minor loop characteristic has been considered in some detail and the DF relating output to input, in the same way as a conventional transfer function, may be graphically constructed (Watton, 1986; Tou and Sculthesis, 1953; Silberberg, 1956). The describing function N_e is considered for the more idealised stiction and coulomb friction with no speed-dependent transition from one to the other. It therefore represents a good approximation to reality, and for this example a stiction/Coulomb friction ratio of 2.0 exists. The DF is of course also input amplitude dependent, the input in this example being hydraulic torque.

13mm lines, 10.73m long

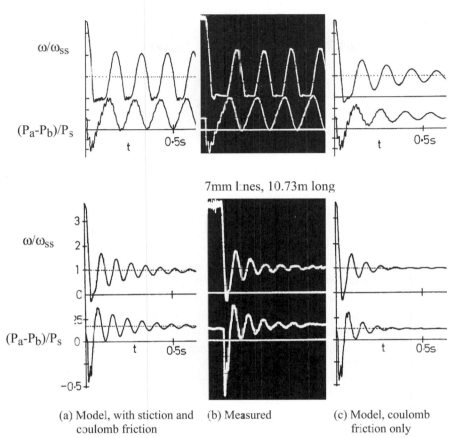

7mm Lnes, 10.73m long

(a) Model, with stiction and   (b) Measured   (c) Model, coulomb
  coulomb friction   friction only

Figure 2.61. Motor speed control in the presence of long lines and motor nonlinear friction

Rearranging Figure 2.63, and including an estimate of the servovalve transfer function G_{sv}, then allows the system characteristic equation to be written:

$$\frac{K_v G_{sv} + T_1}{\frac{sL}{2R_v}\left[T_2 + \frac{2T_1 R_v}{R_m}\right]} = -\frac{1}{N_e}$$

$$\frac{1}{R_m} = \frac{1}{R_i} + \frac{1}{2R_e} \quad K_v = \frac{Ga\,k_i\,H_t}{T_1} \quad L = \frac{J}{D_m^2} \quad R_v = \frac{1}{k_p} \quad (2.185)$$

$$G_{sv} \approx \frac{1}{1 - \frac{2\varsigma s}{\omega_n} + \frac{s^2}{\omega_n^2}}$$

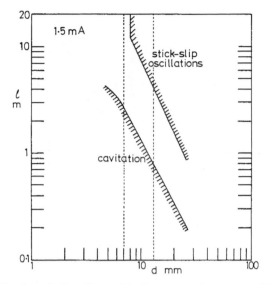

Figure 2.62. Limiting boundaries of operation for an open-loop motor drive with long lines and motor nonlinear friction

Instability occurs when equation (2.185) is satisfied, that is, the frequency domain plot of the left-hand side intersects the DF plot on the right-hand side. In this example the critical intersection point turns out to be the –1 point on the negative real axis, and plots for just two motor speeds are shown in Figure 2.64. It is evident that both line dynamics and servovalve dynamics must be included in the system model in order to predict the condition for instability with reasonable accuracy. It is also evident that a lossless line simplification would have only resulted in a gain error of less than 10%. This could be useful in the sense that the frequency transfer function is more easily evaluated using a lossless line model. Considering a range of motor speeds results in Figure 2.65, which illustrates the variation in the gain to cause closed-loop instability, together with the corresponding frequency of oscillation.

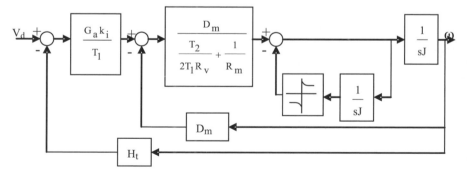

Figure 2.63. Linearised block diagram of a speed control system with long lines

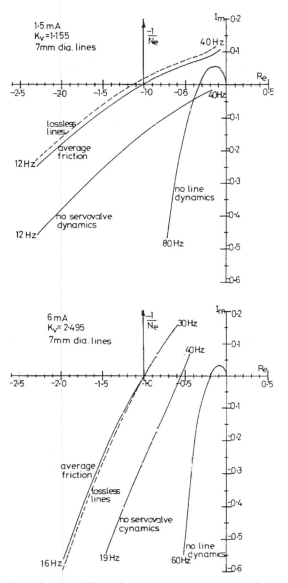

Figure 2.64. Condition for instability of a closed-loop motor speed control system with long lines and nonlinear motor friction, and at two speeds

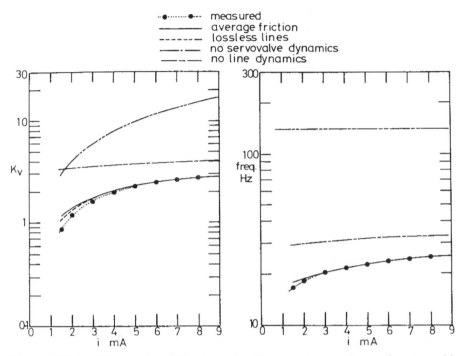

Figure 2.65. Conditions at instability for a closed-loop motor speed control system with long lines and nonlinear motor friction

2.13 Effective Bulk Modulus

From the previous examples it will be clear that fluid compressibility, in conjunction with moving mass effects, is the key feature responsible for hydraulic circuit pressure transients. It is important to know the correct bulk modulus of the fluid and ways in which it can vary, often significantly, from data provided by the manufacturer. Bulk modulus may be defined by the following finite difference form relating the change in volume ΔV compared with the initial volume V and for a pressure change ΔP:

$$\beta = -\frac{\Delta P}{\dfrac{\Delta V}{V}} \tag{2.186}$$

Figure 2.66 shows a typical variation of bulk modulus with pressure and temperature for an ISO 32 mineral oil. The bulk modulus characteristic shown is the result of carefully controlled laboratory tests and in practice oil always contains air in solution as well as possible air bubbles mixing with the oil during working conditions. These additional air contamination aspects can reduce the bulk

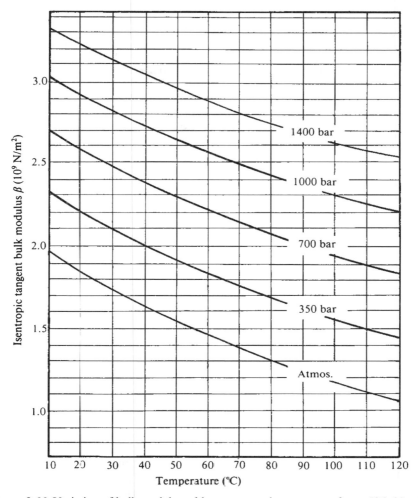

Figure 2.66. Variation of bulk modulus with pressure and temperature for an ISO 32 mineral oil

modulus quite drastically although in practice it is almost impossible to ascertain the effects let alone find an acceptable theory to calculate the change in bulk modulus. The effect of pipe elasticity on bulk modulus is also important particularly if flexible hose is used. In fact this issue of flexible hose is often more significant than the effect of typical air bubbles content.

The effect of different pipe/container volumes can also be included by noting the parallel law for bulk modulus addition following Figure 2.67.

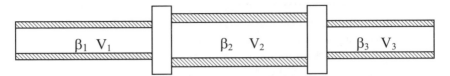

Figure 2.67. A combination of pipes each having different volumes and material properties

The effective bulk modulus β_e for the pipe group is given by

$$\frac{1}{\beta_e} = \frac{V_1}{V\beta_1} + \frac{V_2}{V\beta_2} + \frac{V_3}{V\beta_3} \qquad (2.187)$$

Total volume $V = V_1 + V_2 + V_3$

$$\frac{1}{\beta_1} = \frac{1}{\beta_o} + \frac{1}{\beta_{p1}}$$

$$\frac{1}{\beta_2} = \frac{1}{\beta_o} + \frac{1}{\beta_{p2}} \qquad (2.188)$$

$$\frac{1}{\beta_3} = \frac{1}{\beta_o} + \frac{1}{\beta_{p3}}$$

β_o is the oil bulkmodulus β_p is the pipe bulk modulus

For a steel pipe the material bulk modulus is given by

$$\frac{1}{\beta_p} = \frac{2}{E}\left[\frac{d_o^2 + d_i^2}{d_o^2 - d_i^2} + \sigma\right] \qquad (2.189)$$

where for steel $E = 2 \times 10^{11}\,N/m^2$ and $\sigma = 0.25$. Hence for a steel pipe 13 mm internal diameter with 1.5 mm thick walls its bulk modulus is $19.5 \times 10^9\,N/m^2$ compared with mineral oil typical data from Figure 2.66 of $1.6 \times 10^9\,N/m^2$. The effective bulk modulus of this oil/steel pipe is then $1.49 \times 10^9\,N/m^2$ compared with the oil-only value of $1.6 \times 10^9\,N/m^2$, a reduction of 6.9%.

When considering air in solution and/or in bubble form, an estimate of the effective bulk modulus can be made. Considering the gas laws in conjunction with oil compressibility leads to the following estimate of the effective bulk modulus:

$$\frac{\beta_e}{\beta_o} = \frac{\left(\dfrac{P}{P_o}\right)^n + \alpha}{\dfrac{\alpha\beta_o}{nP} + \left(\dfrac{P}{P_o}\right)^n} \qquad (2.190)$$

where P is the pressure, P_o atmospheric pressure, n the air index of compression, α the air/oil volume ratio. Considering typical parameters gives the characteristic shown as Figure 2.68.

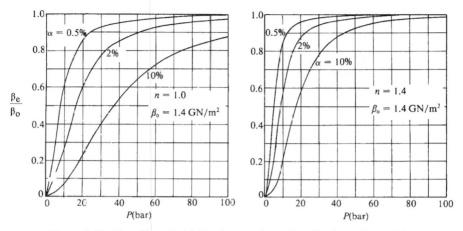

Figure 2.68. The effect of air/oil volume ratio on the effective bulk modulus

For air contamination volumes in practice the effect is probably small at pressures above 100 bar providing air bubble formation is prevented or at least minimised. In practice the effect of small air bubbles in the fluid added to steel elasticity effects means that a good working rule is to reduce the quoted oil bulk modulus by 10%.

The effect of flexible hose elasticity is highly significant when compared with steel piping although manufacturers' data is usually not available in detail. It is possible to experimentally measure the effect for a length of hose by using fast acting flow meters at either end of the line as shown in Figure 2.69.

The particular flow meter used is discussed in Chapter 3 and has been extensively applied to a variety of problems. In this application a mean pressure is set and a transient test performed using a servovalve-controlled flow supply. During the transient test the flow is suddenly changed and the two restrictors are set such that the end pressures vary by ± 10 bar. This means that the effect of pressure on bulk modulus may be neglected. From the flow continuity equation with no moving boundary:

$$Q_i - Q_o = \frac{V}{\beta}\frac{dP}{dt}$$

$$\int_0^t (Q_i - Q_o)\,dt = P_2 - P_1$$

(2.191)

where P_1 and P_2 are the pressure beginning and end-values of the test, the difference being set to 20 bar. In this study, two examples will be presented.

- A rigid steel accumulator-type pressure vessel with a volume of 4.92 litres. This allows determination of the oil bulk modulus only.
- A flexible hose is considered having a nitrile/2-wire mesh/neoprene design, 2.23 m long, internal diameter 20.75 mm. The line volume is therefore $7.54 \times 10^{-4}\,m^3$ with an additional volume of $3.04 \times 10^{-4}\,m^3$ due to the flow meters and fittings. This allows determination of the hose bulk modulus using oil values from the previous tests.

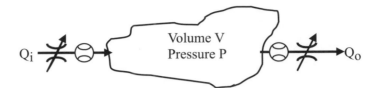

Figure 2.69. Determining effective bulk modulus via dynamic testing

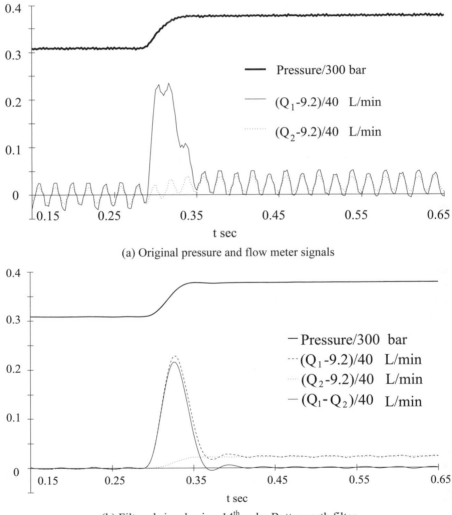

(a) Original pressure and flow meter signals

(b) Filtered signals via a 14th order Butterworth filter

Figure 2.70. Transient test results for bulk modulus determination and using a rigid pressure vessel [Watton and Xue 1994]

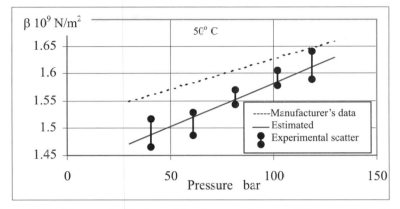

Figure 2.71. Estimate of oil bulk modulus from transient tests [Watton and Xue 1994]

Integration of the flow rates is easily achieved via computer data acquisition although the signals need filtering to reduce noise and pump ripple effects. A 14th-order Butterworth filter was used on all dynamic signals, with a cut-off frequency of 20 Hz set just below the flow ripple frequency measured at 24 Hz (2 × pump speed).

Typical measured transient data are shown in Figure 2.70 for the rigid container test. Using this data and the simple theory outlined earlier, the bulk modulus for the oil only may be calculated and some results are shown in Figure 2.71.

The data from Figure 2.71 is then used to determine the hose bulk modulus by considering the parallel law given by equations (2.187) and (2.188) and a typical result is shown in Figure 2.72.

The deduction from Figure 2.72, that the bulk modulus for hose is significantly lower than that for oil, is well known although the value of typically 30% of that for oil may not be fully appreciated. Clearly in a hose/steel pipe/hose configuration the relative volumes of each component dictate the effective bulk modulus, but the effect of the hose must be included.

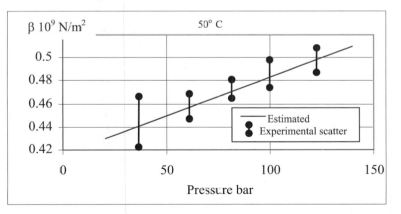

Figure 2.72. Estimated hose bulk modulus from transient tests [Watton and Xue 1994]

2.14 Fluid Viscosity and Density

Where large pressure fluctuations and/or temperature are expected, the affect on viscosity and density could be important. Figures 2.73 and 2.74 show typical data for a mineral oil.

Useful conversion factors are:

$$v = \frac{\mu}{\rho} \qquad (2.192)$$

where v is the kinematic viscosity (m^2/s but often given in centiStokes) and μ is the absolute viscosity (Ns/m^2 but often given in centiPoise).

$$
\begin{aligned}
&1\,cP = 10^{-3}\,Ns/m^2 &&\text{and} &&1\,cS = 10^{-6}\,m^2/s \\
&\mu\,(Ns/m^2) = 10^{-6}\,\rho v &&\text{using} &&\rho\,kg/m^3 \quad\text{and}\quad vcS
\end{aligned}
\qquad (2.193)
$$

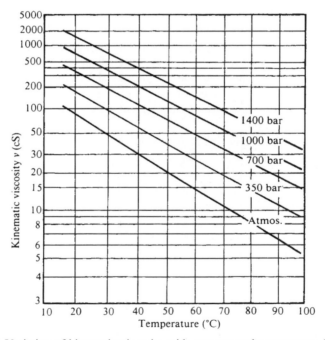

Figure 2.73. Variation of kinematic viscosity with pressure and temperature, ISO 32 mineral oil

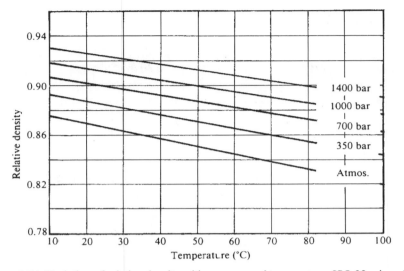

Figure 2.74. Variation of relative density with pressure and temperature, ISO 32 mineral oil

2.15 Solving the System Equations – Simulation Software

To illustrate some basic simulation concepts consider a servovalve/cylinder pressure control system example. Pressure control is found in many applications such as injection moulding and steel mills where high forces have to be applied to the work rolls to ensure that thickness, profile and quality is maintained. Consider the schematic of an injection moulding machine shown in Figure 2.75. Velocity is also measured (or computed on-line) and the control computer switches between velocity and pressure control. However here we will only consider the pressure control phase as shown.

To convey simulation concepts we will make the following assumptions:

- During the pressure control phase, piston movement is negligible, $U \approx 0$
- The volumes on either side of the piston are equal $V_1 = V_2 = V$
- The actuator is double-rod
- Proportional control only will be considered
- Servovalve dynamics will be neglected

Considering the basic equations for this system gives:

$$\left(P_1 - P_2\right)A = P_{load}A + M\frac{dU}{dt} \tag{2.194}$$

$$Q_1 = k_f i\sqrt{P_s - P_1} = AU + \frac{V}{\beta}\frac{dP_1}{dt} \tag{2.195}$$

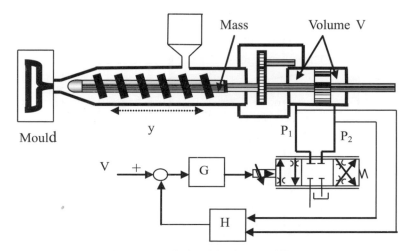

Figure 2.75. Control of an injection moulding process

$$Q_2 = k_f i \sqrt{P_2} = AU - \frac{V}{\beta} \frac{dP_1}{dt} \tag{2.196}$$

Note that dynamically $P_1 + P_2 = P_s$

$$i = G\left[V_d - H_p\left(P_1 - P_2\right)\right] \tag{2.197}$$

These equations can actually be combined and solved by direct integration in the absence of load mass effects. However, this produces a complex, untidy, implicit solution.

Now it is very rare that a direct solution can be obtained for more "realistic" hydraulic control systems, and some form of simulation software package has to be used. There are many such packages available ranging from generalised packages that deal simply with block symbols to purpose-designed packages that allow circuit constructions with icons that look like the hydraulic component they represent. They all use a "click and drag on screen" approach and incorporate a range of numerical integration routines that are automatically implemented once the hydraulic circuit has been assembled by interconnection of the circuit blocks.

Here we will first consider a generalised package, MATLAB Simulink®, that forms just one component of a suite of control and signal processing toolboxes. The Simulink® block diagram is as shown in Figure 2.76 and incorporates the following data:

- Servoamplifier gain $G_a = 2$ mA/V
- Servovalve flow constant $k_f = 2\,e^{-8}$ (flow is m³/s with pressure in Pascals)
- Pressure transducer gain $H_p = 0.05$ V/bar
- Supply pressure $P_s = 100$ bar

- Actuator volume on each side $V = 10\,L$
- Oil/line effective bulk modulus $\beta = 1.4\,e^9\,N/m^2$
- Initial condition for both pressures is half supply pressure $= 50\,bar$
- Demand load pressure of 75 bar is set, the input demand voltage is 3.75 V.

During closed-loop control the pressure P_1 increases to 87.5 bar and the pressure P_2 decreases to 12.5 bar. The pressure differential response is shown in Figure 2.77 and the shape of the response is dependent upon the initial conditior of the pressures.

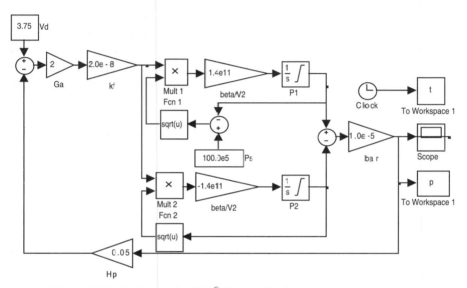

Figure 2.76. MATLAB Simulink® diagram for the pressure control system

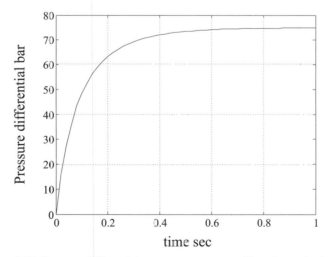

Figure 2.77. Pressure differential step response to a sudden change in demand

MATLAB Simulink® is a very powerful simulation tool, it can generate linearised models for control studies, it can be interfaced with other packages such as solid-body modelling software, it can be interfaced into a real-time control system, and can accommodate typical hydraulic circuits. A particular component model, such as a servovalve, can be grouped into a single block encapsulating the detailed steady-state and dynamic characteristics. One disadvantage is that the graphical appearance of the simulation circuit is not similar to the hydraulic circuit, the feeling of a hydraulic circuit is lost, and this has led to the evolution of a number of custom-developed packages that allow "direct" modelling of typical hydraulic components via icons that look like the components. New models can of course be developed and added to the library. One example of a component oriented simulation package is DSH*plus*. Figure 2.78 shows the same pressure control system assembly of components, including pump and pressure relief valve, using standard library models. Also shown are the transient pressure results similar to those shown in Figure 2.77 with pressure relief valve dynamics neglected.

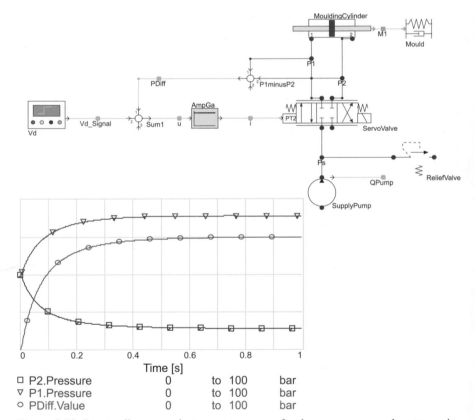

Figure 2.78. System diagram and pressure responses for the pressure control system using DSH*plus*

Design changes or fault conditions are easily incorporated into simulations such as those previously described. Hydraulic simulation software packages minimise the work in setting up the circuit and require no mathematical analysis beyond knowing/assessing/acquiring the important characteristic data for each component. They are continually being improved by the addition of more component models and other modelling techniques such as neural networks. Other highly-developed simulation software packages available are BDSP, Bath*fp*, AMESim, VisSim, HyPneu, DYNAST, ITI-SIM, HOPSAN.

2.16 Transient Response and Stability of a Pressure Rate-controlled Two-stage Pressure Relief Valve

A pressure relief valve is now considered in which the main spindle motion is controlled by applying a back pressure, that is, a pressure to the top end of the main valve spindle. This pressure is controlled by a separate spool valve operating from the same supply pressure that is to be controlled. Using this approach, the pressure rate of increase can be accurately controlled and set to a value that is preferable for large flow rate systems. The concept is shown in Figure 2.79 together with the notation used in the ensuing analysis.

The master valve spool displacement, x, is governed by the supply pressure and this displacement allows fluid to pass through the master valve to the top of the main stage. At the same time the supply pressure is applied to the piston of the main stage causing its displacement, y. This interaction, combined with fluid compressibility effects at the main stage top volume, allows the main pressure to be rate-controllable. The actual valve combination is shown in Figure 2.80.

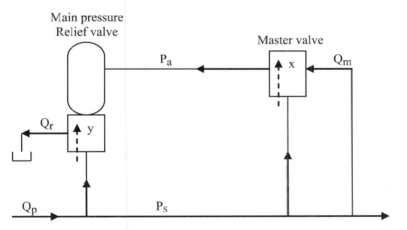

Figure 2.79. Pressure rate controllable pressure relief valve concept (courtesy of Oilgear Towler UK Ltd)

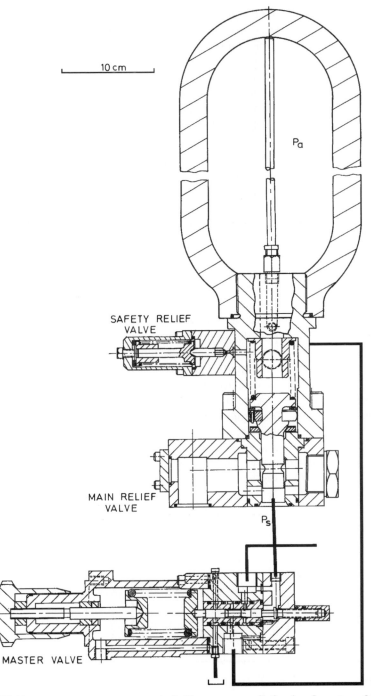

Figure 2.80. Two-stage pressure rate-controllable pressure relief valve [courtesy Oilgear Towler UK Ltd]

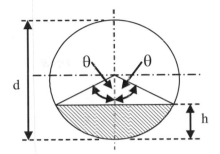

Figure 2.81. Exposed flow area of a circular port

Consider now the dynamic model of the valve system, and assume for testing validation that all the pump flow is directed through the valve with the load shut off. Before the dynamic analysis is pursued we need to realise that the flow control ports are circular. As a piston/spindle/spool type of control element uncovers a circular port then the area exposed is as indicated in Figure 2.81.

$$\frac{a}{a_m} = \frac{\theta}{\pi} - \frac{\sin 2\theta}{2\pi}$$

$$\cos\theta = 1 - \frac{2h}{d} \qquad\qquad (2.198)$$

$$a = \text{area}$$

$$a_m = \text{maximum area } \frac{\pi d^2}{4}$$

The area variation, a, with opening, h, is therefore nonlinear, particularly for small openings typical in valve practice, as shown in Figure 2.82.

Useful approximations to this area variation are as follows:

$$\frac{a}{a_m} \approx 1.7 \left[\frac{h}{d}\right]^{1.5} \qquad \text{with a 5.3\% error at } \frac{h}{d} = 0.15 \qquad (2.199)$$

$$\frac{a}{a_m} \approx 1.7 \left[\frac{h}{d}\right]^{1.5}\left[1 - \frac{0.3h}{d}\right] \quad \text{with a 2.2\% error at } \frac{h}{d} = 0.15 \qquad (2.200)$$

Considering now the dominant equations gives:

$$Q_p = Q_r + a_{sp}\frac{dy}{dt} + Q_m + \frac{V_s}{\beta}\frac{dP_s}{dt} \qquad\qquad (2.201)$$

$$Q_m = -a_{sp}\frac{dy}{dt} + \frac{V_a}{\beta}\frac{dP_a}{dt} \qquad\qquad (2.202)$$

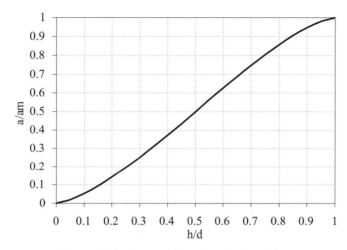

Figure 2.82. Area variation of a circular orifice

$$Q_r = 1.7C_q a_s \left(\frac{y-y_o}{d_s}\right)^{1.5}\sqrt{\frac{2P_s}{\rho}} \qquad\qquad y > y_o \qquad\qquad (2.203)$$

$$Q_m = 1.7C_q a_m \left(\frac{x_i - x}{d_m}\right)^{1.5}\sqrt{\frac{2(P_s - P_a)}{\rho}} \qquad\qquad x < x_i \qquad\qquad (2.204)$$

For the displacements in question, and for circular ports for both the main stage valve and the master valve, the simplest area approximation has been used in the flow equations (2.203) and (2.204). Note that the main stage must move a distance y_o before flow can pass through it, and also the master valve may have an initial displacement x_i.

The system volume is V_s, the main stage top volume is V_a, the main spindle cross-sectional area is a_{sp}, the main spindle radial cross-sectional-area is a_s, the master valve radial cross-sectional area is a_m.

The effect of master valve spool displacement on flow continuity is considered negligible. If this is not the case for other applications then the dynamic force equation for the master valve spool must be included. However, the dynamic force equation for the main stage spool is needed and the equations for both the main stage spindle and the master stage spool are assumed to be as follows:

$$(P_s - P_a)a_{sp} = k_v y + B^*\frac{dy}{dt} + F_r + m\frac{d^2 y}{dt^2} \qquad\qquad (2.205)$$

$$(P_s - P_{mo})a_m = kx \qquad\qquad P_s > P_{mo} \qquad\qquad (2.206)$$

$$F_r = 2C_q^2 \cos\theta\, 1.7a_s\left(\frac{y-y_o}{d_s}\right)^{1.5} P_s \qquad\qquad y > y_o \qquad\qquad (2.207)$$

where, θ is the jet angle at the main stage spindle flow exit area, a_m is the master valve spool cross-sectional-area, m is the main spindle mass, k_v is the main valve spring stiffness, and F_r is the main valve spindle flow reaction force.

The main spool has a directional damper with damping coefficient, B^*, changing with the sign of velocity. This provides more viscous damping when the spindle is closing. The master valve spring stiffness is k, and P_{mo} is the main spindle initial equivalent force setting.

It is sometimes useful to write a set of equations in non-dimensional form since it can indicate the relative significance of some terms and allows transient data to be compared on the same scale relative to steady-state values. It is also necessary to use this approach if a simulation package such as MATLAB Simulink® is used to analyse the behaviour.

In this example the steady-state supply pressure and flow rate are 207 bar and 238 litres/min. The reference values used to non-dimensionalise are twice the steady-state values for all time-varying variables. Also a non-dimensional time scale can be useful where the simulation run time $\tau = \lambda t$. These transformations then lead to the following simulation equations.

These equations form a typical stiff set of equations often met in fluid power systems simulation. In this example the main relief valve spindle response is very fast when compared with the rise in each pressure. However, this does not present a difficult simulation solution issue and a comparison between simulation and measurement is shown in Figure 2.83.

A particularly useful addition to this simulation work was that extremely good pressure rate control was shown to be possible with a much reduced top volume and hence much reduced cost.

$$\frac{d\overline{P}_s}{d\tau} = \alpha_1 \left[\overline{Q}_p - \overline{Q}_r - \alpha_2 \frac{d\overline{y}}{d\tau} - \overline{Q}_m \right] \tag{2.208}$$

$$\overline{Q}_r = \alpha_3 \left(\overline{y} - \overline{y}_o \right)^{1.5} \sqrt{\overline{P}_s} \qquad \overline{y} > \overline{y}_o \tag{2.209}$$

$$\overline{Q}_m = \alpha_4 \left(\overline{x}_i - \overline{x} \right)^{1.5} \sqrt{\overline{P}_s - \overline{P}_a} \qquad \overline{x} < \overline{x}_i \tag{2.210}$$

$$\frac{d\overline{P}_a}{d\tau} = \alpha_5 \left[\overline{Q}_m + \alpha_2 \frac{d\overline{y}}{dt} \right] \tag{2.211}$$

$$\overline{x} = \overline{\alpha}_6 \left[\overline{P}_s - \overline{P}_{mo} \right] \qquad \overline{P}_s > \overline{P}_{mo} \tag{2.212}$$

$$\frac{d^2\overline{y}}{d\tau^2} = \alpha_7 \left[\overline{P}_s - \overline{P}_a - \alpha_8 \overline{y} - \alpha_9 \frac{d\overline{y}}{d\tau} - \overline{F}_r \right] \tag{2.213}$$

$$\overline{F}_r = \alpha_{10} \left(\overline{y} - \overline{y}_o \right) \overline{P}_s \qquad \overline{y} > \overline{y}_o \tag{2.214}$$

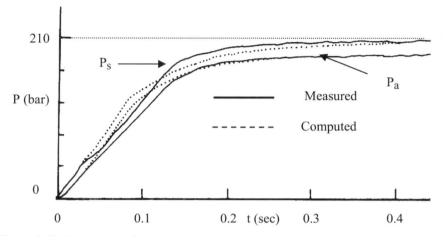

Figure 2.83. Comparison of computed and measured pressure responses for a pressure rate-controllable pressure relief valve

The parameters for the valve assembly are as follows:

$\alpha_1 = 0.18$ $\alpha_2 = 0.057$ $\alpha_3 = 2.24$ $\alpha_4 = 0.819$

$\alpha_5 = 1.07$ $\alpha_6 = 2.67$ $\alpha_7 = 945$ $\alpha_8 = 0.01$

$\alpha_9 = 0.12$ for positive velocity, 0.4 for negative velocity

System volume $V_s = 33.9$ litres, main valve top volume $V_a = 6.17$ litres

$\lambda = 50$

Consider next the stability of this pressure relief valve. The dynamic equation set is highly nonlinear due to the flow equations and the bi-directional damper characteristic for the main stage spindle. Assuming the usual techniques of linearisation developed earlier, and assuming also a mean damper coefficient results in the following equations:

$$\delta Q_p = \delta Q_r + a_{sp} \frac{d\,\delta y}{dt} + \delta Q_m + \frac{V_s}{\beta} \frac{d\,\delta P_s}{dt} \tag{2.215}$$

$$\delta Q_m = -a_{sp} \frac{d\,\delta y}{dt} + \frac{V_a}{\beta} \frac{d\,\delta P_a}{dt} \tag{2.216}$$

$$\delta Q_r = k_r \delta y + \frac{\delta P_s}{R_r} \tag{2.217}$$

$$\delta Q_m = \frac{\delta P_s}{R_{mv}} - \frac{\delta P_a}{R_{mv}} - k_{mv} \delta y \tag{2.218}$$

$$\left(\delta P_s - \delta P_a\right)a_{sp} = k_v\delta y + B^*\frac{d\,\delta y}{dt} + \delta F_r + m\frac{d^2\delta y}{dt^2} \tag{2.219}$$

$$\delta P_s a_m = k\,\delta x \tag{2.220}$$

$$\delta F_r = k_1\delta P_s + k_2\delta y \tag{2.221}$$

Now consider the block diagram Figure 2.84.

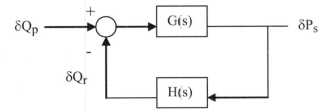

Figure 2.84. Block diagram of the pressure control concept, linearised analysis

Considering the open-loop transfer function from Figure 2.84 and combining the linearised equations shows that the valve can be unstable (Watton, 1990). However, stability will be helped by first ensuring that the following condition is satisfied:

$$k_{mv} = \frac{1}{R_{mv}} = 0 \tag{2.222}$$

In other words, circular ports rather than rectangular ports must be used, exactly what the valve manufacturer has done. Hence for this master valve design, incorporating (2.222), the open-loop linearised transfer function becomes:

$$G(\overline{s})H(\overline{s}) = \frac{\alpha + \varepsilon + \overline{R}\,\overline{s} + X\overline{s}^2}{\overline{s}\left(1 + \overline{V}\alpha + \overline{V}\overline{R}\,\overline{s} + \overline{V}X\overline{s}^2\right)} \tag{2.223}$$

$$\overline{R} = \frac{R_B}{R_r} \qquad R_B = \frac{B^*}{a_{sp}^2} \qquad R_r = \frac{2P_s}{Q_{r\,\text{steady-state}}} \qquad \overline{V} = \frac{V_s}{V_a}$$

$$\alpha = 1 + \frac{V_a\left(k_v + k_2\right)}{\beta a_{sp}^2} \qquad \varepsilon = \frac{3P_s V_a}{\beta h a_{sp}} \qquad X = \frac{m\beta}{V_a a_{sp}^2 R_r^2}$$

$$h = \text{main stage spindle opening} \qquad \overline{s} = s\frac{V_a R_r}{\beta}$$

Stability of the third-order characteristic equation is achieved providing:

$$\overline{R}\left(1 + \alpha\overline{V} + \overline{R}\right) > X\left(\alpha + \varepsilon\right) \tag{2.224}$$

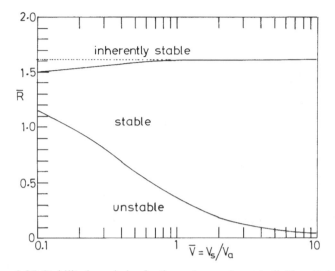

Figure 2.85. Stability boundaries for the pressure rate-controllable relief valve

Considering the data for this valve, the stability regions may be plotted as shown in Figure 2.85. Recalling that the volume ratio $\overline{V} = 5.49$ for the experimental test condition, then the ratio $\overline{R}$ must be greater than typically 0.1 for stability. Considering actual values results in:

$$\frac{B^*}{a_{sp}^2 R_r} = 1.52 \text{ for positive velocities}$$

$$\frac{B^*}{a_{sp}^2 R_r} = 5.71 \text{ for negative velocities}$$

(2.225)

Clearly the valve operation is stable as identified from the transient test shown in Figure 2.83. In fact the damping characteristics designed means that the valve is almost at the inherently stable condition, which from Figure 2.85 occurs when $\overline{R} = 1.61$.

2.17 Data-based Dynamic Modelling of Components – Accumulator Charging Behaviour

Reflecting on the previous examples concerned with mathematical modelling it will be quickly realised that many parameters in the model are difficult to determine, such as velocity damping coefficients and flow reaction forces, and inevitably some form of experimental testing will be called for. It may be extremely difficult and sometimes impossible to accurately determine some parameters so the question could be asked: why not use input/output measured data to generate

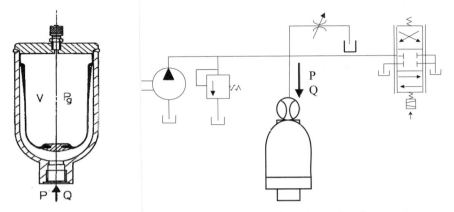

Figure 2.86. Dynamic testing of an accumulator using a fast acting flow meter

the component model including the unknown data? This may not be acceptable as a design aid for the component but it could allow system simulation to progress more rapidly. If the dynamic performance of the component is to be analysed then a knowledge of its dynamic equations, even if in linearised form, could be extremely helpful in obtaining a suitable structure for the data-based model.

This example considers a nitrogen gas filled accumulator of the bladder type and dynamic modelling requires a knowledge of the complex dynamic gas charging characteristic together with some estimate of the entrance pressure/flow characteristic. During operation it is a matter of debate whether the accumulator gas law should be modelled as isothermal or reversible adiathermal. It is also known that the gas law exponent $PV^n = \text{constant}$ in excess of $n = \gamma = 1.4$ is used by accumulator manufacturers for some dynamic design calculations.

If we therefore consider the technique using a fast acting flow meter, as discussed earlier for estimating bulk modulus, then during accumulator operation the oil flow rate in or out may be dynamically measured. This allows the gas law to be used to determine the dynamic interpretation. The method therefore uses measured data combined with a pragmatic theoretical model allowing the gas charging/discharging characteristic to be determined as a function of time and consequently as a function of input/output flow rate. The test method is shown in Figure 2.86.

A servovalve is used to rapidly switch flow into and out of the accumulator and the flow rate and pressure are measured via computer. Inevitably for hydraulic measurements, the signals have to be filtered and in this case an 8th-order Chebyshev filter was used.

Typical raw data for one test are shown in Figure 2.87. The filtered data for 1,2,4 litre accumulators are then used to evaluate the gas "constant" under dynamic conditions. Application of the mass flow continuity equation for flow into the accumulator gives:

$$\left(P-P_g\right)a_b + \frac{\rho Q^2}{a_n} - \frac{\rho Q^2}{a_b} = \frac{\left(m_o + \rho \int Qdt\right)}{a_b}\frac{dQ}{dt} \qquad (2.226)$$

a_n is the cross-sectional area of the neck
a_b is the cross-sectional area of the bladder
m_o is the initial mass of oil in the accumulator
P_g is the gas pressure

All the terms in (2.226) may be numerically evaluated from the data, but the result of many calculations show that for this study all flow terms on the left-hand side are negligible.

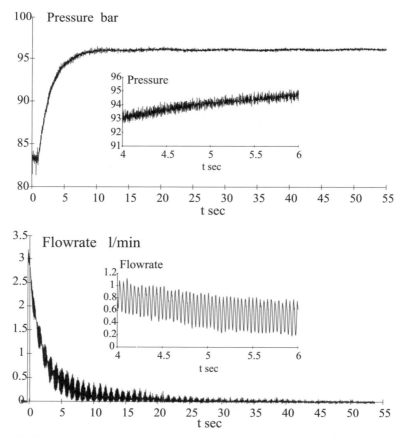

Figure 2.87. Transient pressure and flow rate data for a 1 litre accumulator charging [Watton and Xue 1995]

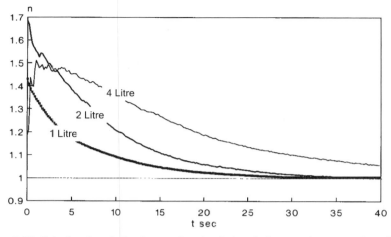

Figure 2.88. Calculated variation in gas charging index during transient operation of accumulators [Watton and Xue 1995]

Considering then the gas law:

$$PV^n = \text{constant} = P_0 V_0^n \qquad (2.227)$$

$$n = \frac{\log P / P_0}{\log V_0 / V} \qquad (2.228)$$

$$V = V_0 - \int Q dt \qquad (2.229)$$

Computed values of the index n may then be determined and the variation with time is shown in Figure 2.88 for three accumulator sizes. It can be seen that the initial rapid charging gives rise to an index greater than the adiabatic value $n = 1.4$ and this settles down to the isothermal case as time progresses and the charging process slows down.

Results for both charging and discharging can then be plotted against flow rate to give a useful design aid to system simulation. These results are shown in Figure 2.89. From these results it is deduced that:

$$n = 1 + K_a \left(1 - e^{-Q/Q_a}\right) \qquad (2.230)$$

and the constants K_a and Q_a are shown in Table 2.4.

Table 2.4. Experimentally determined constants for determining the charging exponent n

Accumulator volume (litres)	K_a	Q_a
1	0.45	0.92
2	0.70	1.87
4	0.54	1.42

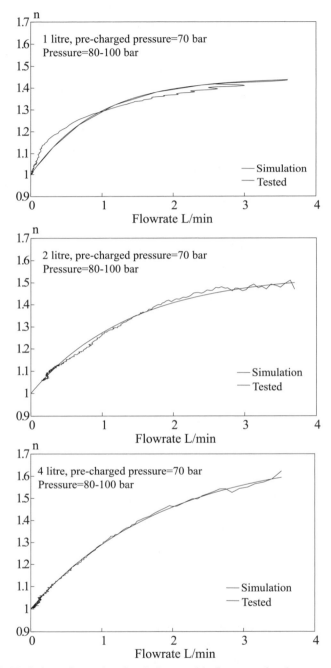

Figure 2.89. Variation of gas charging index n with flow rate for three accumulators of volume 1,2 4 litres [Watton and Xue 1995]

2.18 Data-based Dynamic Modelling of Components – Time Series Analysis

Now consider the application of time series analysis using transient pressure and flow rate data obtained in a manner similar to the accumulator previously discussed. However, in this case, as in most time series analysis cases met by the author, we use a linearised dynamic analysis of the component to aid the establishment of the time series structure.

Consider then a single-stage type with a novel nonlinear damper that aids opening yet offers a greater resistance to closing. This characteristic produces a very stable valve performance. The valve is tested using the same setup as for the previous accumulator test, and is shown in Figure 2.90.

Considering the mathematical model of the valve, using equations similar to those discussed for the proportional pressure relief valve in Section 2.11, leads to two nonlinear terms due to the valve pressure/flow characteristic and the nonlinear damper. Linearising these equations and assuming a mean damping coefficient for the nonlinear damper, results in the following linearised transfer function, effectively the dynamic impedance:

$$\frac{\delta P(s)}{\delta Q(s)} = Z(s) = R_v \frac{\left(1 + c_1 s + c_2 s^2\right)}{1 + d_1 s + d_2 s^2} \tag{2.231}$$

The steady-state valve linearised "resistance" R_v and the transfer function constants are given by:

$$R_v = \frac{1}{\dfrac{k_x a_s}{k} + k_p} \qquad c_1 = \frac{B}{k} \qquad c_2 = \frac{m}{k}$$

$$d_1 = \frac{a_s^2 + k_F B}{k_x a_s - k_p k} \qquad d_2 = \frac{m k_p}{k_x a_s + k_p k} \tag{2.232}$$

$$k_x = \frac{\partial Q}{\partial x} = \frac{1.5 Q_{ss}}{x_{ss}} \qquad k_p = \frac{\partial Q}{\partial P} = \frac{Q_{ss}}{P_{ss} - P_b}$$

P_{ss} and Q_{ss} are the valve working steady-state pressure, 70 bar, and mean flow rate 22.5 l/min

P_b is the back pressure, 9 bar

m = moving mass, 0.34 kg

B = mean damper coefficient, 0.3×10^4 N/ms⁻

k = spring stiffness, 34.3×10^3 N/m

a_s = spindle cross sectional area, 63.6×10^{-6} m²

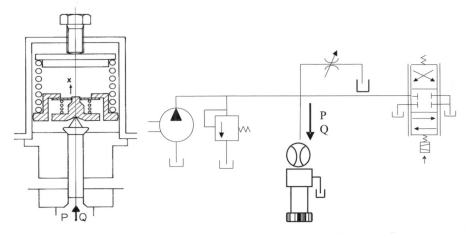

Figure 2.90. Dynamic testing of a pressure relief valve using a fast acting flow meter

To determine the time series structure we observe that the dynamic impedance is second order. A suitable Z transformed approximation is therefore:

$$\frac{\delta P(z)}{\delta Q(z)} = Z_v\left(z^{-1}\right) = \frac{b_0 + b_1 z^{-1} + b_2 z^{-2}}{1 - a_1 z^{-1} - a_2 z^{-2}} \qquad (2.233)$$

Arranging the sampled time series then gives:

$$\begin{aligned}
\delta P(kT) &= a_1 \delta P\left((k-1)T\right) + a_2 \delta P\left((k-2)T\right) \\
&\quad + b_0 \delta Q(kT) + b_1 \delta Q\left((k-1)T\right) + b_2 \delta Q\left((k-2)T\right)
\end{aligned} \qquad (2.234)$$

The least square method is then used by considering sufficient data sets, each set being selected from one sample shift of the transient pre-filtered pressure and flow rate data. In this study a sampling frequency of 1 kHz was used. Transforming (2.234) into matrix notation gives:

$$y = A\beta \qquad (2.235)$$

and the least squares solution for the unknown coefficients is then given by:

$$\beta = \left(A^T A\right)^{-1} A^T y \qquad (2.236)$$

$$
\begin{bmatrix}
\delta P(3T) \\
\delta P(4T) \\
\delta P(5T) \\
\vdots \\
\delta P(nT)
\end{bmatrix}
=
\begin{bmatrix}
\delta P(2T) & \delta P(nT) & \delta Q(3T) & \delta Q(2T) & \delta Q(T) \\
\delta P(3T) & \delta P(2T) & \delta Q(4T) & \delta Q(3T) & \delta Q(2T) \\
\delta P(4T) & \delta P(3T) & \delta Q(5T) & \delta Q(4T) & \delta Q(3T) \\
\vdots & \vdots & \vdots & \vdots & \vdots \\
\delta P((n-1)T) & \delta P((n-2)T) & \delta Q(nT) & \delta Q((n-1)T) & \delta Q((n-2)T)
\end{bmatrix}
\begin{bmatrix}
a_1 \\
a_2 \\
b_0 \\
b_1 \\
b_2
\end{bmatrix}
$$

$$\quad\; y \qquad\qquad\qquad\qquad\qquad\qquad\qquad A \qquad\qquad\qquad\qquad\qquad\qquad\qquad \beta$$

$$(2.237)$$

Transient test data are shown in Figure 2.91 again illustrating the importance of pre-filtering the data prior to further analysis. Applying the least squares identification method to this data then gives the estimated coefficients. Here it was found that convergence occurred after typically 100 data sets and showing negligible change up to 300 data sets. The coefficients are:

$$b_o = 1.619 \times 10^{10} \qquad b_1 = -2.972 \times 10^{10} \qquad b_2 = 1.359 \times 10^{10}$$

$$a_1 = 1.883 \qquad a_2 = -0.922 \tag{2.238}$$

A measure of the accuracy of the method can be obtained by considering the magnitude of the low frequency and high frequency asymptotes of the two transfer functions in the s domain and the z domain, equations (2.231) and (2.233):

$|Z_v|$ s domain		$|Z_v|$ z domain	

$$R_v = \cfrac{1}{\cfrac{k_x a_s}{k} + k_p} = 0.14 \times 10^{10} \qquad \cfrac{b_o + b_1 + b_2}{1 - a_1 - a_2} = 0.15 \times 10^{10} \tag{2.239}$$

$$\frac{R_v c_2}{d_2} = \cfrac{\cfrac{k_x a_s}{kk_p} + 1}{\cfrac{k_x a_s}{k} + k_p} = 1.64 \times 10^{10} \qquad b_o = 1.62 \times 10^{10} \tag{2.240}$$

The comparisons are extremely good which may also be judged by the predicted flow rates for the data shown in Figure 2.91.

By using a suitable transformation from the z domain to the s domain, the valve impedance frequency response may be determined. The choice of transformation is crucial and in this example the simple backward difference transformation and the bilinear transformation did not produce good results at the higher frequencies. A second-order transformation, however, did give good results using the substitution as follows:

$$z^{-1} = \frac{1 - sT/2 + s^2 T^2/8}{1 + sT/2 + s^2 T^2/8} \tag{2.241}$$

Magnitude and phase comparisons, via the further substitution $s = j\omega$ are shown in Figure 2.92.

The comparisons shown in Figure 2.92 are excellent, indicating a phase lead impedance characteristic. The time series analysis approach has shown to be useful even in the presence of a significant damping nonlinearity.

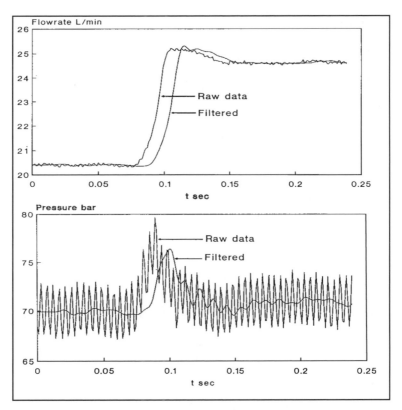

(a) transient pressure and flow rate data

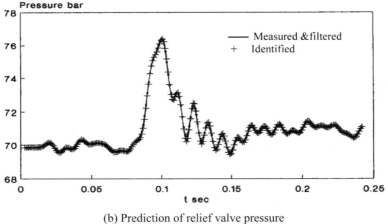

(b) Prediction of relief valve pressure

Figure 2.91. Time series modelling [Watton and Xue 1995]

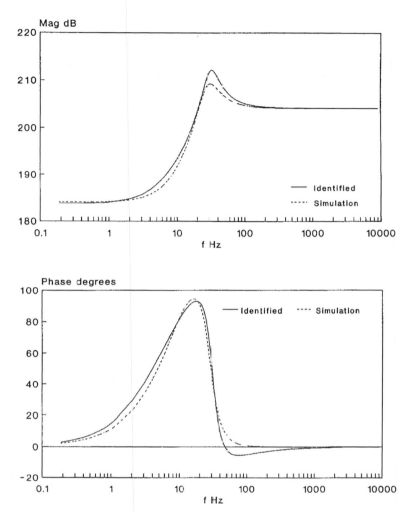

Figure 2.92. Computed impedance frequency response from a time series analysis [Watton and Xue 1995]

2.19 Data-based Dynamic Modelling of Components – Long Lines

It has been shown that the modelling of transmission lines is complex for a single steel line let alone a combination of lines and flexible hose, some perhaps of different diameters. One way of determining the dynamic model for a combination of lines is to use a data-based approach. It can be applied either in situ or to a laboratory test circuit built from the same line elements, and fast acting flow meters are required at each end of the line assembly. A similar approach was used for the experimental determination of bulk modulus for lumped elements as discussed in

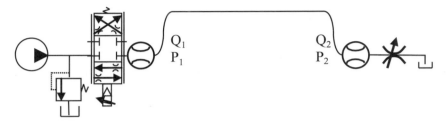

Figure 2.93. Measuring the transient pressures and flow rates at the ends of a long line

Sections 2.13 and 2.17 although in this application the transient data are analysed in more detail. Consider therefore Figure 2.93.

If the hyperbolic form of the transmission line equations are considered then it can be shown that sampled data transformations relating sums and differences of pressures and flows may be written as follows:

$$\frac{(P_1 + P_2)}{2(Q_1 - Q_2)} = A(z^{-1}) = \frac{a_0 + a_1 z^{-1} + ... + a_i z^{-i}}{1 + b_1 z^{-1} + ... + b_j z^{-j}} \quad i = n, j = n+1 \tag{2.242}$$

$$\frac{(P_1 - P_2)}{2(Q_1 + Q_2)} = B(z^{-1}) = \frac{a_0 + a_1 z^{-1} + ... + a_i z^{-i}}{1 + b_1 z^{-1} + ... + b_j z^{-j}} \quad i = n-1, j = n \tag{2.243}$$

where n = number of modes selected, 2, 4, 5, etc.

It is also important to note that the poles of $A(z^{-1})$ represent the even-numbered modes and the poles of $B(z^{-1})$ represent the odd-numbered modes.

The unknown coefficients may be evaluated using the least squares method. Rearranging (2.242) and (2.243) then results in equations suitable for the least squares method as follows:

$$P_s(kT) = 2\sum_{i=0}^{n} a_i Q_d\left((k-i)T\right) - \sum_{j=1}^{n+1} b_i P_s\left((k-i)T\right) \tag{2.244}$$

$$P_d(kT) = 2\sum_{i=0}^{n-1} a_i Q_s\left((k-j)T\right) - \sum_{j=1}^{n} b_i P_d\left((k-j)T\right) \tag{2.245}$$

T = sampling interval

It should be obvious from the background theory that $B(z^{-1})$ embodies an integrator term which is mapped to the pole z = 1. Hence one test for the coefficient estimate accuracy is that $1 + \Sigma b_i = 0$. In addition it can be shown that both line resistance and effective bulk modulus may be estimated from the $A(z^{-1})$ and $B(z^{-1})$ functions as follows:

$$R = \left(\frac{128\mu L}{\pi d^4} \text{ for a uniform line}\right) \rightarrow 4B(z^{-1})\Big]_{z=1} \tag{2.246}$$

$$\beta_e = \left(1 - z^{-1}\right)\frac{V}{T}A(z^{-1})\Big]_{z=1} \tag{2.247}$$

The method is particularly useful for line combinations where transmission line theory becomes extremely difficult to handle mathematically. Consider the details of just one line tested:

- Steel line 13 mm internal diameter 6.3 m long, with 3 m long flexible hose either end, each hose having a nominal diameter of 10 mm.
- Line inlet temperature 50°C, exit temperature 41°C.
- Fluid viscosity $v = 30$ cS @40°C, 20 cS @50°C.
- Fluid bulk modulus $\beta = 1.5\,e^9\,N/m^2$.
- Sampling frequency 1 kHz, 250 samples used in least squares.
- Validation tests must be done at a different load restrictor settings.

Figure 2.94 shows predicted and actual pressure transients using $n = 2,6$ modes indicating good comparisons. The pressures are evaluated using the measured flow data to compute $(P_1 - P_2)$ and $(P_1 + P_2)$. Table 2.5 shows the estimated coefficients for $A(z^{-1})$ and $B(z^{-1})$.

Table 2.5. Coefficients estimated using the method of least squares

$A(z^{-1})$							
$a_i/10^9$	$n=2$	$n=4$	$n=5$	b_i	$n=2$	$n=4$	
a_0	3.0961	2.1963	2.0463	1	1.0000	1.0000	1.0000
a_1	−4.3085	−0.5784	1.1258	b_1	−1.3534	−1.3043	−1.2043
a_2	1.5312	0.2737	−0.5110	b_2	0.2747	0.1312	0.0936
a_3		−4.4589	−7.6039	b_3	0.0788	0.3156	0.2404
a_4		3.0889	5.0538	b_4		0.0951	−0.0080
a_5			2.5196	b_5		−0.2374	0.0421
a_6			−1.8385	b_6			0.0678
				b_7			−0.2317
				Sum	0.0001	0.0001	−0.0001

$B(z^{-1})$							
$a_i/10^9$	$n=2$	$n=4$	$n=6$	b_i	$n=2$	$n=4$	
a_0	1.1493	0.5144	0.9160	1	1.0000	1.0000	1.0000
a_1	−1.1436	0.9359	1.0526	b_1	−1.8883	−1.7098	−1.6194
a_2		−1.7038	−1.7034	b_2	0.9387	0.6760	0.6037
a_3		0.2617	−0.2097	b_3		0.0051	0.1828
a_4			−0.5334	b_4		0.0978	−0.1143
a_5			0.4918	b_5			−0.1978
				b_6			0.2541

Table 2.6. Experimentally obtained modal frequencies and their comparison with modal theory for three cases

Line 1 Steel 7 mm dia, 11.8 m long		Line 2 Steel 13 mm dia 12.8 m long		Line 3 Hose-Steel-Hose 12.3 m long
Frequency (Hz)		Frequency (Hz)		Frequency (Hz)
Measured	Theory	Measured	Theory	Measured
n				
1 46.7	51.5	47.3	47.5	36.1
2 89.0	104.6	94.5	95.7	81.9
3 162.9	157.8	157.5	144.0	130.4
4 211.0	211.2	193.3	192.4	176.8
5 267.0	264.6	265.0	240.8	194.7
6 316.0	318.1	304.0	289.8	279.3

Figure 2.94 shows a very good prediction using n = 6 modes and it will also be seen from Table 2.5 that the integrator pole at z = 1 is accurately identified from $A(z^{-1})$ even using n = 2 modes. To determine the modal frequencies, a second-order matched Z transform technique is used to map from the z domain to the s domain s follows:

$$\left(z_r \pm jz_i\right) \rightarrow z^2 - 2ze^{-aT}\cos bT + e^{-2aT} \rightarrow \left(s + a \pm jb\right) \qquad (2.248)$$

Therefore evaluating the poles of $A(z^{-1})$ and $B(z^{-1})$ from Table 2.5 enables the modal frequencies to be calculated from (2.248) although n = 10 modes were needed to produce sufficiently accurate results.

The experimentally derived frequencies are shown in Table 2.6 together with experimental and theoretical results for two other steel lines, one 7 mm internal diameter and 11.8 m long, the other 13 mm internal diameter and 12.8 m long.

Table 2.7. Experimentally obtained bulk modulus and resistance from dynamic data

n	Line 1 $\beta / 10^9 \, \mathrm{Nm}^{-2}$	Line 2 $\beta / 10^9 \, \mathrm{Nm}^{-2}$	Line 3 $\beta / 10^9 \, \mathrm{Nm}^{-2}$
2	1.23	1.38	0.73
4	1.25	1.40	0.75
6	1.25	1.40	0.75
	Average 1.24	Average 1.39	Average 0.74
n	Line 1 $R / 10^9 \, \mathrm{Nm}^{-2}$	Line 2 $R / 10^9 \, \mathrm{Nm}^{-2}$	Line 3 $R / 10^9 \, \mathrm{Nm}^{-2}$
2	3.11	0.10	0.45
4	3.12	0.13	0.48
6	3.16	0.16	0.51
	Average 3.13 Measured 2.90	Average 0.13 Measured 0.30	Average 0.48 Measured 0.47

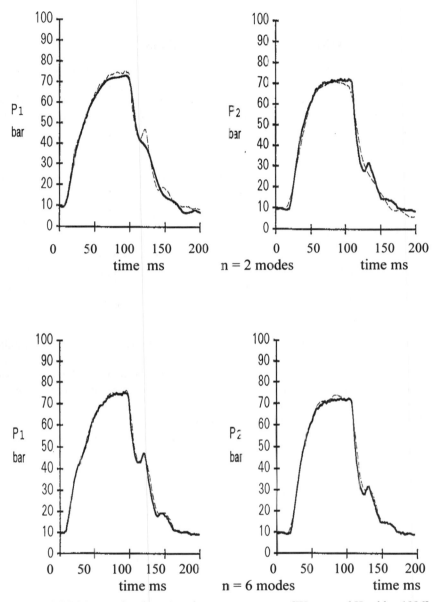

Figure 2.94. Measured and computed pressure responses [Watton and Hawkley 1996]

Estimates of line resistance and effective bulk modulus using (2.246) and (2.247) are shown in Table 2.7 up to n = 6 modes.

The comparisons shown in Table 2.6 and Table 2.7 are good and give confidence to the approach which is based on a number of experimental and mathematical steps each prone to discussions of appropriateness and confidence error.

The reduction in effective bulk modulus and hence modal frequencies is again evident for a line combination including flexible hose. Line resistance estimation is rather poor for the 13 mm diameter steel line but this is due to the accuracy of the pressure transducers used given such small pressure differences existing for line 2 in comparison with the other two lines.

The data-based approach used here could have some value as a line condition monitoring method. However, its value for circuit modelling and hence as a model-based system condition monitoring aid may be more significant.

2.20 Data-based Dynamic Modelling of Components – Artificial Neural Networks (ANNs)

2.20.1 Introduction

ANNs for engineering applications have evolved from biological and neurophysiological concepts. It is estimated that the human brain contains a complex interconnection net of $10^{10} \rightarrow 10^{11}$ neurons or nerve cells. A simplified schematic of a biological neuron is shown in Figure 2.95.

There are typically $10^{3} \rightarrow 10^{4}$ dendrites (inputs) per neuron and these dendrites connect the neuron to other neurons. They either receive inputs from other neurons, via contacts called synapses, or connect other dendrites to the synaptic outputs. The synapses are capable of changing a dendrite's local potential in a positive or negative direction and can be either excitatory or inhibitory in accordance with the ability to strengthen or dampen the neuron excitation. The storing of

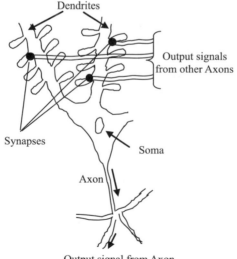

Figure 2.95. Schematic of a biological neuron

information in a neuron is thought to be concentrated at the synaptic connections, in particular the pattern of these connections and strengths (weights) of the synaptic connections. Human synapses are of a complex chemical nature whereas synapses in nervous systems of primitive animals, such as insects, are predominantly based on electrical signal transmission. Assuming a simplified model of the neuron, the cell body receives inputs from other neurons through adjustable or adaptive synaptic connections to the dendrites. The output signal from a cell is transmitted along a branching axon to the synapses of other neurons. When a neuron is excited it produces a train of pulses which are transmitted along an axon to the synaptic connections of other neurons. The output pulse rate depends upon both the strength of the input signals and the weights, or strength, of the corresponding synaptic connections. The maximum firing rate is 1000 pulses per second and is controlled by the excitatory synapse which increases the pulse rate and the inhibitory synapse which reduces the pulse rate or even blocks the output signal. We therefore have the concept of an activation function. It has been discovered that for many biological neurons a linear summation approximation is appropriate and the neuron's output is proportional, in some range, to a linear combination of the neuron's input signal values. We therefore have the concept of a neuron "firing" via the activation function when particular conditions of information occur at the input.

2.20.2 The Artificial Neuron – Forming the ANN

The biological neuron has received a great deal of attention, but the basic mathematical model takes a very simplified yet powerful form as shown symbolically in Figure 2.96.

Figure 2.96 illustrates a summation of input signal from other neurons, each signal having a weight attached to it. For mathematical reasons a bias signal is also included to simulate equation constants. The summation of weighted signals then passes through the activation function before being transmitted to other neurons in the neural network representing the engineering system. This is developed further in Figure 2.97. In engineering practice, we have to consider a collection of many neurons to create the Artificial Neural Network (ANN) to represent the

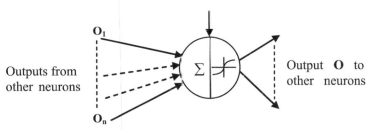

Figure 2.96. Basic neuron representation

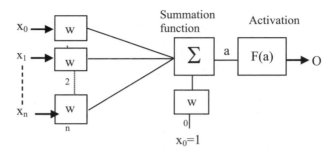

Figure 2.97. Single neuron model

application, although not in the millions as is the biological case. A topology is usually adopted as shown in Figure 2.98, which represents just one of many possible ANN topologies.

The inputs to the ANN are the measured data corresponding to known conditions, both good and faulty, and the output/s are the corresponding condition indicators. There are usually $1 \rightarrow 2$ hidden layers, and if we are trying to predict one fault then clearly there will be one output neuron indicating true or false. Figure 2.98 can accommodate at least 2 faults, assuming both can occur, and assuming the notation $-1 =$ no fault and $1 =$ fault then:

	Output 1	Output 2
No Faults	-1	-1
Fault A and Fault B	1	1
Fault A only	1	-1
Fault B only	-1	1

Once a topology has been selected a training scheme must be applied to adjust the unknown neuron weights and biases such that the input data maps to the assigned fault condition output data. In some simple applications this mapping can be exact, as shown in the following example of an exclusive OR gate. The possible inputs and outputs are:

Inputs	x_1	0 0 1 1
	x_2	0 1 0 1
Output		0 1 1 0

We next select a suitable activation function and the most common are shown in Figure 2.99.

Since this is a binary logic problem it makes sense to use a unipolar switching activation function $0 \rightarrow 1$ and a suitable ANN topology is shown in Figure 2.100 for a particular evaluation with $x_1 = 0$ and $x_1 = 1$.

It will be deduced from Figure 2.100 that all the input data pairs are exactly mapped to the required fault indicators of either 0 or 1.

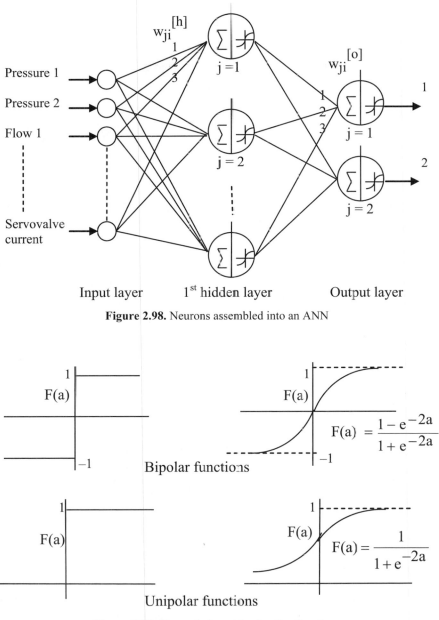

Figure 2.98. Neurons assembled into an ANN

$$F(a) = \frac{1 - e^{-2a}}{1 + e^{-2a}}$$

Bipolar functions

$$F(a) = \frac{1}{1 + e^{-2a}}$$

Unipolar functions

Figure 2.99. Step and sigmoid activation functions

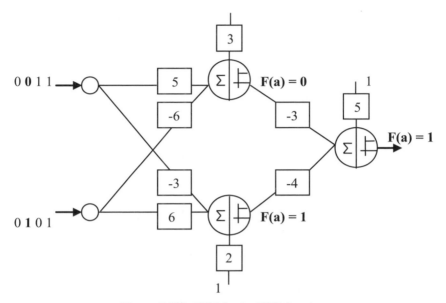

Figure 2.100. ANN for the XOR function

2.20.3 Training ANNs – The Basic Weight Change Equation for a Single Neuron

Having selected a network topology (and there are very few absolute rules for this), we now have to determine the network weights at each neuron such that the output of the network reflects the condition represented by the input data. In this study we wish the output/outputs to represent a fault/faults condition. Usually the input data will be a sequence of data that could be steady-state measurements or time-varying measurements of one or more parameters. The method that forms the foundation of ANN training is called the backpropagation method. The method revolves around minimising the (sum of error)^2 between the network desired output, d, and the actual output during training, y. This is no more than the method we use to fit data using the method of least squares. The error function is summed for all the input patterns, p, used. The weights are changed such that E achieves the desired value set by the user. It does not follow that this is an absolute minimum value since often it continually, but slowly, reduces with training time. In this case training is stopped when there is evidence of little return for the extra time involved, again a typical engineering judgement in practice. The backpropagation method adjusts the neuron weights using the delta rule as follows:

- *For the output neuron*

This is valid for the output layer but the local error cannot be evaluated for the hidden layers. Generally the local error for a hidden layer is evaluated with the knowledge of the local error at the upper layer. Thus δ at the upper layer can be propagated backwards to the hidden layer.

$$\Delta w_{ji}^{[o]} = \eta \, \delta_j^{[o]} \, o_i^{[h]}$$

$$\delta_j^{[o]} = \left[d_{jp} - y_{jp} \right] \left[\frac{dF(a)}{da} \right]_j^o \tag{2.249}$$

- *For the hidden layer*

$$\Delta w_{ji}^{[h]} = \eta \, \delta_j^{[h]} \, x_i$$

$$\delta_j^{[h]} = \left[\frac{dF(a)}{da} \right]_j^h \sum \delta_i^{[o]} w_{ij}^{[o]} \tag{2.250}$$

All these terms may be calculated for each data set providing the activation function derivative can be evaluated. This is easily achieved if the activation function is a continuous mathematical function such as the sigmoid functions shown in Figure 2.99. The *learning rate* $0 < \eta < 1$ and must be set by the user, and will typically be very small to aid convergence.

For the unipolar sigmoid function:

$$F(a) = \frac{1}{1 + e^{-2a}} \quad \text{so} \quad \frac{dF(a)}{da} = 2F(a)\left(1 - F(a)\right) \tag{2.251}$$

For the bipolar sigmoid function:

$$F(a) = \frac{1 - e^{-2a}}{1 + e^{-2a}} \quad \text{so} \quad \frac{dF(a)}{da} = 1 - F(a)^2 \tag{2.252}$$

Training times can be long, varying between minutes and several hours, depending on the problem. In practice other mathematical mechanisms are added to improve both training times and convergence. One particularly common addition is the use of *momentum* whereby the weight changes are modified as follows

$$\Delta w_{ji}^{[r]}(k) = \eta \, \delta_j^{[o]} \, o_i^{[r-1]} + \alpha \, \Delta w_{ji}^{[r]}(k-1)$$

$$w_{ji}^{[r]}(k+1) = w_{ji}^{[r]}(k) + \Delta w_{ji}^{[r]}(k) \tag{2.253}$$

$$0 < \alpha < 1$$

ANNs have been studied extensively in recent years and there are other techniques to aid training and also a variety of ANN types. These may be pursued via the literature, but the author has found that the backpropagation method previously outline has worked well in many practical applications including hydraulic systems fault diagnosis, dynamic modelling, feedback adaptive control.

2.20.4 Some Practical Issues

- The inputs are usually in vector form.
- The first hidden layer will contain more neurons than the number of inputs.
- The second layer, should it be needed, could have more/less neurons than the first layer but probably less. There are no specific rules for layer or neuron selection.
- Neurons are usually scaled between $0 \rightarrow 1$ or $-1 \rightarrow +1$ depending on the type of problem being studied.
- The use of ± 1 outputs tends to give good training discrimination between the no-fault and fault conditions.
- Training will not produce exactly the ± 1 outputs and some "fuzzy boundaries" will have to be adopted, for example:

    $-1 < \text{no fault} < -0.8$

    $0.8 < \text{fault} < 1$

- The activation function $F(a)$ is selected to match the problem being studied. For fault diagnostic problems, and many other engineering problems such as dynamic modelling, a sigmoid-type function is readily suited to the technique required for training the neural network to fit the measured data.
- ANN software programs, for example MATLAB® ANN Toolbox, will require the use of network training constants selected from experience.
- ANN topology construction requires trial and error regarding both the number of neurons and the number of hidden layers. Typically, the first hidden layer will contain more neurons than the number of inputs. Any subsequent layers then tend to reduce the number of neurons leading to the 'diamond' rule.
- Trained ANNs should be validated using different training data sets, often termed 'unseen data'.
- Very low learning rates and momentum may be required, resulting in long training times.
- ANNs will usually only predict using new data within, or near to, the data range used to train the network.

2.20.5 Modelling a Long Line in a Pressure Control System

Consider the basic pressure control system shown in Figure 2.101 with a rigid load such that changes in actuator volume may be neglected under pressure control. A pressure transducer is used for feedback control which is achieved via a servovalve. Two fast acting flow meters are positioned at the ends of each line allowing dynamic signals to be used for line dynamics identification.

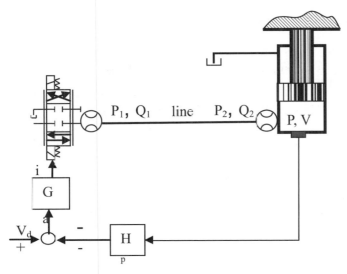

Figure 2.101. Pressure control system considered as a fixed load volume

Recall that the transmission line equations for the lossless case produced a pair of equations, which are repeated here:

$$P_2 + Z_{ca}Q_2 = z^{-1}\left(P_1 + Z_{ca}Q_1\right) \tag{2.254}$$

$$P_1 - Z_{ca}Q_1 = z^{-1}\left(P_2 - Z_{ca}Q_2\right) \tag{2.255}$$

$$Z_{ca} = \text{lossless line characteristic impedance } \sqrt{\frac{\rho\beta}{a^2}} \tag{2.256}$$

Had the distributed line model been considered then this pair of equations would be modified by the complex impedance terms as follows:

$$P_2 + F_1 Z_{ca}Q_2 = z^{-1}F_2\left(P_1 + F_1 Z_{ca}Q_1\right) \tag{2.257}$$

$$P_1 - F_1 Z_{ca}Q_1 = z^{-1}F_2\left(P_2 - F_1 Z_{ca}Q_2\right) \tag{2.258}$$

F_1 and F_2 are modified Bessel functions, as may be determined from Section 2.12, but they need not be pursued since the objective here is to utilise the power of an ANN to map these functions via time domain training. In addition we note that for a fixed volume load we can make a sampled data transformation relating Q_2 to P_2, for example the following backward difference form. This means that we can use either of the transmission line equations to reduce the identification of Q_2 using measurements of Q_1, P_1, P_2 via the following functional Φ:

$$Q_2 = \Phi(z^{-1}P_1 + z^{-1}P_2 + z^{-1}Q_1) \tag{2.259}$$

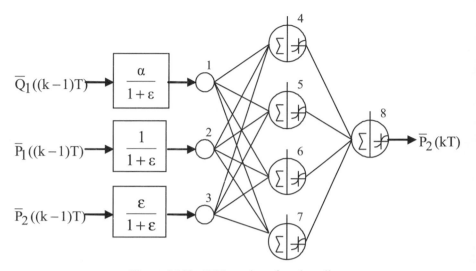

Figure 2.102. ANN topology for a long line

This time relationship can been achieved quite simply by using a three-input ANN with one hidden layer containing four neurons. Also to keep the data within the ±1 range the pressures and flow rates are referenced to maximum values and scaling constants are introduced such that the ANN appears as shown in Figure 2.102. This approach also means that the flow meter measurement at the end of the line is not needed.

Consider the following data:

  Steel line length $L = 14$ m, diameter $= 13$ mm

  Fluid density $\rho = 860$ kg/m^3, bulk modulus $\beta = 1.43 \times 10^9$ N/m^2

  $C_o = 1289$ m/s, $T = L/C_o = 0.0109$ s

  Pressure reference $P_{ref} = 100$ bar, flow reference $Q_{ref} = 6$ L/min

  $R_{ref} = P_{ref}/Q_{ref} = 10^{11}$ Ns/m^5, $\tau = V R_{ref}/\beta = 0.079$ s

  Line characteristic impedance $Z_{ca} = 8.35 \times 10^9$ Ns/m^5, $\alpha = Z_{ca}/R_{ref} = 0.0835$,
  $\varepsilon = \alpha\tau/T = 0.605$

From various trials, select ANN learning rate $= 0.001$, error goal $= 0.0001$, 200 data batches with 1750 samples per batch, to give the result shown in Figure 2.103 which illustrates the transient response of P_2 to a demanded step change from $0 \rightarrow 40$ bar. Note that the ANN requires inputs delayed by $T = 10.9$ ms and so a data acquisition sampling frequency of 1.75 kHz is selected to give 19 samples per delay time T.

The weights and biases are shown in Table 2.8.

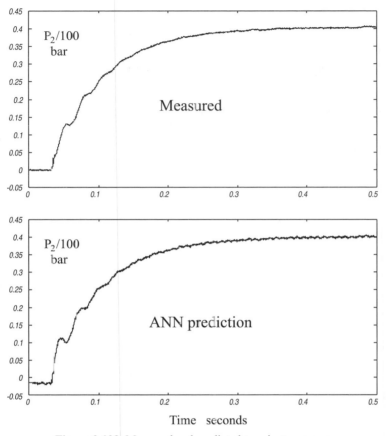

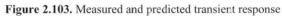

Figure 2.103. Measured and predicted transient response

Table 2.8. Weights and biases for a 3-4-1 ANN transmission line topology

$w_{14} = -0.558$			
$w_{15} = -0.914$			
$w_{16} = 0.394$			
$w_{17} = 0.436$			
$w_{24} = 0.884$	$w_{48} = 0.105$	$b_4 = -0.946$	
$w_{25} = -0.247$	$w_{58} = -0.248$	$b_5 = -0.162$	$b_8 = 0.434$
$w_{26} = 0.181$	$w_{68} = 0.674$	$b_6 = -0.664$	
$w_{27} = 0.931$	$w_{78} = 1.075$	$b_7 = -0.019$	
$w_{34} = -0.920$			
$w_{35} = -0.910$			
$w_{36} = 0.161$			
$w_{37} = 0.543$			

Table 2.8 shows no dominant path through the topology although it will be noticed that the bias b_7 has a very small value in comparison with the other weights and biases. It will be shown in Chapter 3 that this bias changes significantly, in comparison to the rest, in the presence of line faults.

2.21 Data-based Dynamic Modelling of Components – the Group Method of Data Handling (GMDH)

2.21.1 Introduction

This method utilises an ANN-type topology based on an assumed dynamic characteristic but without neurons as classically defined for an ANN. For a GMDH network:

- Algebraic polynomial functions relating 'neuron' connections are established and the method of least squares analysis is used to determine the data goodness of fit. These functions can be nonlinear in comparison to the usual linear functions used in ANNs.
- The neurons have no activation function and hidden layers are added to ensure that the demanded goodness of fit is met. In practice a simple minimum root-mean-square error (rms) requirement is used to determine the appropriate topology.
- The path through the topology is then selected by discarding all the other connections such that the minimum rms error path is maintained for that particular hidden layer deemed sufficient.
- The method essentially seeks the "best combination" of dynamic parameters, rejecting those that do not match the goodness of fit requirement.

2.21.2 An Introductory Example – a Servovalve/Motor Drive

For example, consider the servovalve/motor drive shown in Figure 2.104.

For this example a mathematical system model is first used where it is known that the response to a sudden change in servovalve voltage gives an exponential-type speed response due to servovalve damping. The damping of course varies with motor speed for this nonlinear speed control system However, for generality and to ensure the embracement of any higher order dynamics we will assume up to fourth-order dynamics relating current speed to previous samples of both speed and applied voltage as shown in Figure 2.105.

The functional assumed, f, has to be resolved by trial and error and in this case the simplest 2-parameter form was found suitable:

$$f = w_0 x_i + w_1 x_i^2 + w_2 x_i x_j + w_3 x_j^2 + w_4 x_j + w_5 \tag{2.260}$$

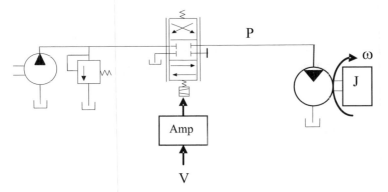

Figure 2.104. Unidirectional servovalve/motor drive

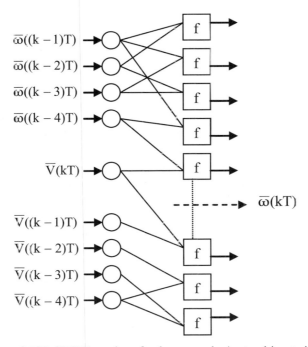

Figure 2.105. GMDH topology for the servovalve/motor drive, training

It will also be deduced that if n inputs are combined r at a time then the number of combinations, hence neurons in the next hidden layer, is given by:

$$\text{number of neurons} = C_r^n = \frac{n(n-1)(n-2)\ldots\ldots(n-r+1)}{r!} \qquad (2.261)$$

In the present study $n=9$ and $r=2$ giving 35 neurons in the first hidden layer. If all 36 outputs from each neuron in the first hidden layer are then combined $r=2$ at a time to provide inputs into a second hidden layer then the number of neurons in the second hidden layer would be 630 and thus prohibitive. Clearly the number of neurons in successive hidden layers must be resricted and this is achieved again by trial and error.

In the present study a data sampling interval $T = 10\,ms$ was selected and 200 data points were used. The mathematical model assumes no motor leakage and the combination of the servovalve square root flow characteristic combined with motor load inertia leads to the following nonlinear equation:

$$\overline{\omega}^2 = \overline{V}^2 \left(1 - \alpha \frac{d\overline{\omega}}{dt} \right)$$

$$\overline{\omega} = \frac{\omega}{\omega_{max}} \qquad \overline{V} = \frac{V}{V_{max}} \qquad \alpha = \frac{J\omega_{max}}{D_m P_s}$$

(2.262)

The parameter α was chosen as 0.03 seconds, various inputs were applied to the system and only one hidden layer was used, training ceasing when an rms error value of 0.005 was first achieved at any of the 36 neuron outputs. The six unknowns in equation (2.260) are calculated using the least squares method with 200 data points at each neuron and then recalculated at the remaining neurons for all of the 36 combinations. At each analysis the predicted waveform is compared with the simulation data and the nett rms error determined.

It perhaps comes as little surprise that the output neuron which first established an rms error below 0.005 was that relating the speed current output to its value at the previous sample together with a measure of the current value of the input voltage. Thus this single output neuron only is retained resulting in the trained network shown in Figure 2.106.

The mimum rms error of 0.00368 resulted from the following weights:

w0 = 0.401

w1 = –0.127

w2 = 0.412

w3 = –0.283

w4 = 0.595

w5 = 0.000148

(2.263)

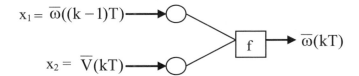

$x_1 = \overline{\omega}((k-1)T)$

$x_2 = \overline{V}(kT)$

f → $\overline{\omega}(kT)$

Figure 2.106. Trained GMDH network for a basic servovalve/motor drive

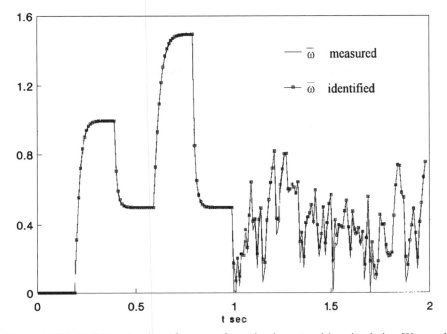

Figure 2.107. GMDH network performance for a simple motor drive simulation [Xue and Watton 1995]

The network has been trained to emulate a nonlinear hydraulic circuit from step response and random signal response tests as shown in Figure 2.107, and validation tests have also been carried out using sinusoidal inputs.

This example shows that given good data and a suitable topology and algorithm, a GMDH approach can be used to model dynamic data for a nonlinear system and can identify the probable causal links between the output and network inputs provided all sensible combinations are considered at the outset. This approach is also referred to as a self-organising ANN approach using the GMDH algorithm.

2.21.3 Application to a Practical Circuit – a Multivariable Problem

Now consider a practical realisation of the previous example, the motor having a proportional pressure relief valve load as shown in Figure 2.108.

The servovalve/motor drive to be modelled has three inputs, supply pressure P, voltage to the servovalve V_d, voltage to the load valve V_L. The data sampling frequency was selected as 500 Hz giving a sampling interval T = 2 ms. To capture

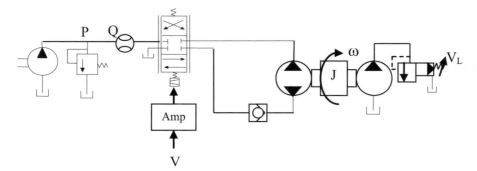

Figure 2.108. Unidirectional servovalve/motor drive with loading device

assumed dynamics the following states are considered for the discrete GMDH approach:

$$\omega((k-1)T), \; \omega((k-2)T), \; \omega((k-3)T)$$

$$P((k-1)T), \; P((k-2)T), \; P((k-3)T), \; P((k-4)T)$$

$$V_d((k-1)T), \; V_d((k-2)T), \; V_d((k-3)T), \; V_d((k-4)T)$$

$$V_L((k-9)T), \; V_L((k-10)T), \; V_L((k-11)T), \; V_L((k-12)T) \qquad (2.264)$$

It will be seen that the known inherent pure delay of typically 18 ms (9T) in the electromagnetic circuit of the load valve is captured by using appropriate delayed sampled states of V_L. It is also anticipated that a third-order model for speed, a fourth-order model for pressure and a fourth-order model for servovalve voltage changes are sufficient to embrace the appropriate dynamics.

The algorithm used, following some alternative trials, is as follows:

$$f = \sum_{i=1}^{6} w_i x_i + \sum_{i=1}^{6} \sum_{j=1}^{6} w_{ij} x_i x_j + w_c \qquad (2.265)$$

A single hidden layer was found best since the mean rms error could not be improved using a second hidden layer, the minimum rms error being 30 rpm for motor speed training and 37 rpm over a wide range of validation conditions. This compares with measured speed peak-to-peak fluctuations of typically 100 rpm. The trained network showed the minimum rms error for the output neuron (one of the original 105) connected to just 6 of the 15 inputs as shown in Figure 2.109. Also shown is the network as used for computer simulation applications.

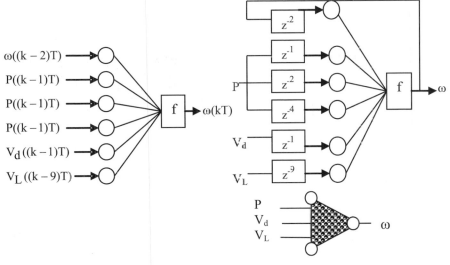

(a) trained network   (b) network for computer simulation

Figure 2.109. GMDH network for a servovalve/motor drive with loading

It is also a straightforward matter to derive a GMDH network with flow as the output enabling the most preferable circuit to be used within a more complex circuit simulation – integration of flow rates is not necessarily required using such circuits providing of course compressibility effects are incorporated into a network or a separate compressibility network is derived. It will be seen from Figure 2.109 that the pure delay characteristic of the load valve has been captured by the analysis and only one sample of the servovalve voltage is required.

A motor drive suffers from significant speed fluctuations due to the motor displacement ripple effect, its frequency of course also proportional to motor speed. Hence for training ANNs or GMDH networks speed measurements must be filtered. In this study a fourth-order Butterworth filter was used. During training, 3000 data patterns were used and some results are shown in Figure 2.110 for a combination of servovalve and load valve voltage changes. Needless to say, the trained network has been validated for other combinations of unseen input signal changes.

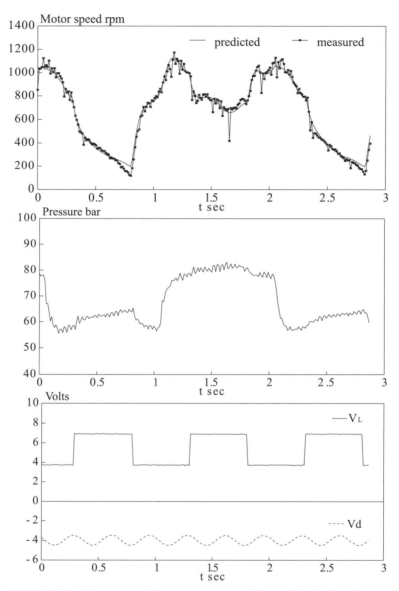

Figure 2.110. Predicted motor speed response to servovalve and load valve changes using a GMDH network [Xue and Watton 1997]

2.21.4 Circuit Simulation Using Interconnected Artificial Networks

Here we specifically consider GMDH sub-models linked together to determine a more detailed performance of a hydraulic circuit by combining the previous motor drive with the other system components necessary for the drive to operate.

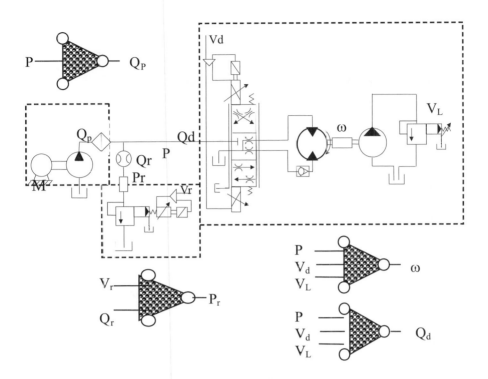

Figure 2.111. Hydraulic circuit and its sub-models in GMDH network form

In addition an ANN is used to establish the simulation initial conditions obtained experimentally. The circuit incorporates a proportional flow control valve and two different proportional pressure relief valves as shown in Figure 2.111. The pump/filter and integral line volume dynamic model is also added to produce four sub-models linked together by flow continuity with negligible compressibility effects. Note that 2 sub-models are now required for the motor drive, one for speed and one for flow rate. In addition, it is practically necessary to insert a flow meter into the pressure relief valve circuit to both obtain its pressure/flow characteristic and observe practical performance, particularly initial conditions. The flow meter must also be included in the circuit simulation model to account for the small yet measurable pressure drop during operation.

The simulation circuit is shown in Figure 2.112 including the flow meter which is retained in the circuit during dynamic testing. Note that the network for the flow meter is no more than its measured steady-state pressure drop/flow characteristic. Also, each sub-model is executed using time step intervals determined by the data acquisition sampling frequency. Different sampling frequencies can be accommodated for each sub-model providing they are integer values.

To enable a more efficient use of the simulation two ANNs are used to model the system initial conditions for a wide range of speeds and control signals. One is used to determine the pressure P and other is used to determine the flow rate Q_d

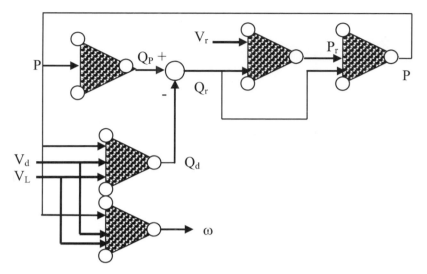

Figure 2.112. System simulation block diagram using GMDH artificial network sub-models

for a range of speeds, ω, load voltage, V_L, and servovalve voltage, V_d. It was found that a 3-5-3-1 topology in both cases was sufficient resulting in a rms error of 0.56 bar for pressure, over a $56 \rightarrow 90$ bar range, and a rms error of 0.27 L/min for flow rate, over a $16 \rightarrow 45$ L/min range.

The ANN approach to modelling the pressure relief valve steady-state characteristic was found to be particularly helpful since the flow rate/pressure slope is nonlinear and different at each cracking pressure setting. The mean rms flow error was found to be 0.86 L/min over a 50 L/min range and significantly better than 3.82 L/min using a least squares analysis with a fixed curve relationship at each setting.

Some results are shown in Figure 2.113(a) and (b).

For the input conditions shown in Figure 2.113, the motor speed prediction is well-representative of that measured. The speed fluctuates between typically $60 \rightarrow 80$ bar but note that the relief valve flow rate fluctuates rapidly to the extent that it suggests a potential wear issue for this system operating mode. The power of a good simulation is the ability to predict, hopefully improve, the circuit behaviour by the introduction of new components or perhaps small modifications to components. The latter aspect is difficult to implement using the current technique in the sense that each sub-model requires data-based training. However, to illustrate the point of component addition here, an increased volume will be introduced. A lumped volume of 1.125 litres is added and its dynamic performance, including connection restriction effects, has been modelled again using a GMDH network. This resulted in a simple network as shown in Figure 2.114(a) together with its insertion into the circuit as shown in Figure 2.114(b).

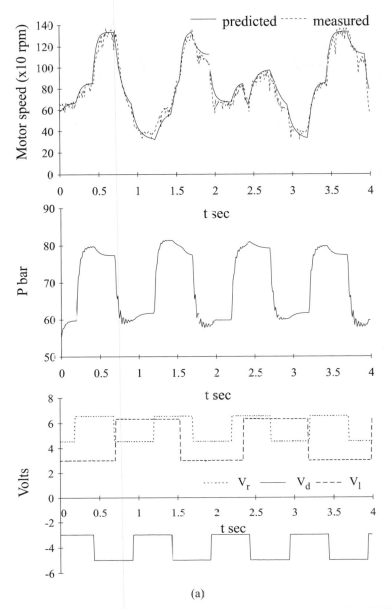

(a)

Figure 2.113. Hydraulic circuit performance prediction using interconnected GMDH artificial network sub-models

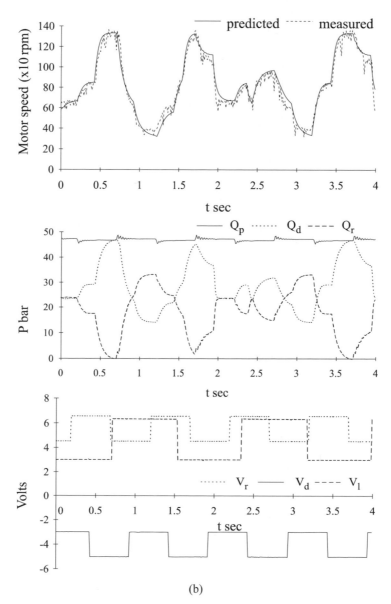

(b)

Figure 2.113. (continued)

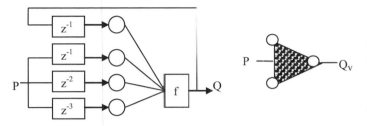

(a) Lumped volume trained GMDH network

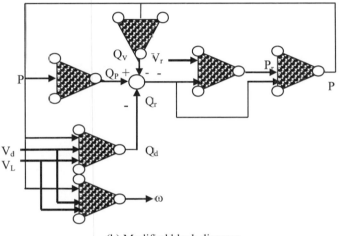

(b) Modified block diagram

Figure 2.114. System simulation with additional volume component

Running this modified simulation then gives the results shown in Figure 2.115(a) and Figure 2.115(b) for similar input conditions to those shown in Figure 2.113 without the added volume.

It may be seen that the added volume has created a slight increase in pressure fluctuation amplitude but the flow fluctuation through the pressure relief valve has been significantly reduced due to the increased compressibility flow rate.

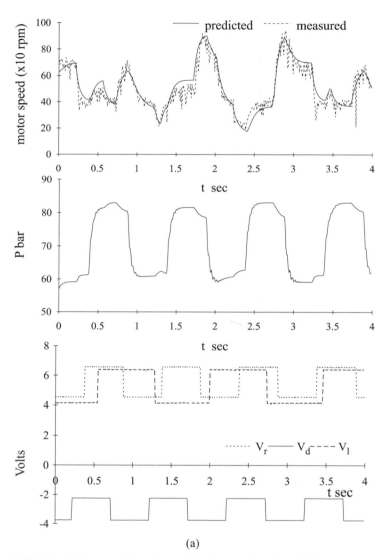

(a)

Figure 2.115. Simulation prediction for hydraulic circuit with added volume and using GMDH sub-models

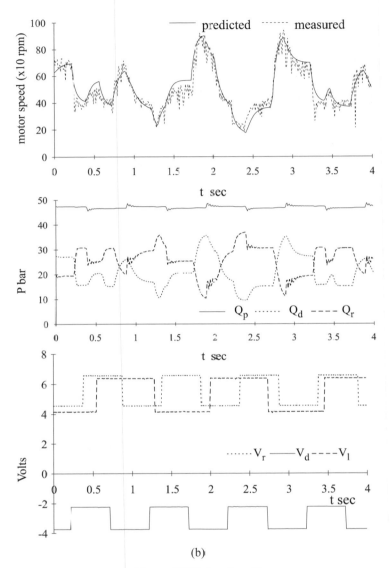

(b)

Figure 2.115. (continued)

2.22 Intelligent Control, Improving Closed-loop Performance

Although this chapter is not intended as an overview of control systems it is appropriate to mention there are many on-line methods now well-developed to improve closed-loop performance, loosely called adaptive control techniques. For example:

- A fuzzy logic approach can be used whereby the type of control action used is based on intelligent rules developed from experience. The type of control action can be set, for example, from the magnitude of the error and the magnitude of the velocity co-state. Since no clear boundaries can be set in practice then fuzzy boundary concepts can be used to decide the most probable control action required.
- A neural network can also be used to adapt its weights to changing conditions and ensure that the closed-loop response obeys a pre-requisite form, as close as practically possible. Also a combination of fuzzy logic and a neural network can been used to improve closed-loop performance.
- Sliding mode control can be applied to hydraulic control systems and employs essentially state feedback and a switching strategy in the error loop. This approach attempts to drive the response across the pre-determined plane defined by the state feedback equation with of course restrictions on the dynamic behaviour possible.
- Predictive control can be used to develop the error signal using the predicted required state at each sampling interval. This requires an accurate model of the hydraulic nonlinear dynamics which may be considerably helped using a neural network model obtained from practical tests.

These methods, and other variations, become useful when the dynamic behaviour of the control system changes. A change can be caused, for example, by different load conditions or different demand conditions. The most nonlinear hydraulic control system is probably a servovalve controlled system due to the non-linear pressure/flow characteristic of the servovalve. This was discussed in Section 2 and it should be deduced that the effect is most noticeable, and undesirable, for a speed control system. Given commercially-available programmable digital control cards, it is possible to avoid the application of sensitive adaptive techniques briefly mentioned earlier. This can be done by ostensibly cancelling the square root pressure/flow characteristic by a software programming approach.

Consider the practical realisation of this "intelligent" control approach to a proportional valve controlled motor as shown in Figure 2.116.

A uni-directional motor is used with a similar design of pump and a proportional pressure relief valve to create load pressure changes. The main principles employed are:

- Supply pressure can be varied using a proportional pressure relief valve
- Load pressure is measured

- A Moog M2000 digital Programmable Servo Controller is used to:
  - (i) compute the proportional control valve pressure/flow characteristic, invert it and pre-multiply the voltage to the valve
  - (ii) generate a dynamic pressure filter to produce a desired dynamic characteristic
  - (iii) adjust the proportional pressure relief valve setting to optimise efficiency
- No speed feedback is used with intelligent control

The control valve measure characteristic is of the following form:

Measured $$Q_1 = (aV - b)\sqrt{P_s - P_1}$$ (2.266)

Required $$Q_1 = a\overline{V}\sqrt{P_{ref}} \rightarrow V = \overline{V}\sqrt{\frac{P_{ref}}{P_s - P_1}} + \frac{b}{a}$$ (2.267)

The linearised control valve practical characteristic is shown in Figure 2.117 where it can be seen that the method is very effective for a control valve differential pressure across the inlet port greater than 15 bar.

Next consider the drive efficiency defined by the measured ratio of (motor output power)/(pump delivered power) and shown as Figure 2.118.

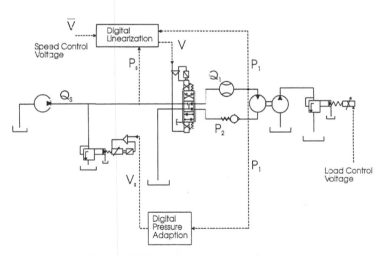

Figure 2.116. Intelligent control of a motor

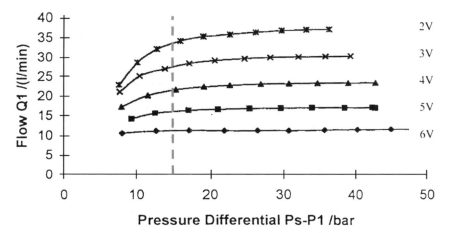

Figure 2.117. Proportional control valve flow linearisation

It can be seen that the efficiency is low, as expected for this type of system, but can be maintained reasonably close to its maximum for proportional flow control valve pressure differentials between 10 bar and 20 bar. Comparing Figures 2.117 and 2.118 suggests that selecting 20 bar pressure drop across the proportional flow control valve should significantly improve the system performance.

Measured characteristic $P_s = -P_{so} + k_s V_s$

but $P_s - P_l = 20\,bar$ $\qquad V_s = \dfrac{P_l + 20 + P_{so}}{k_s}$ $\qquad\qquad$ (2.268)

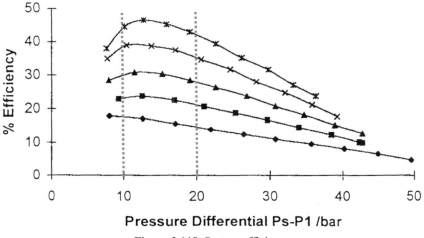

Figure 2.118. System efficiency

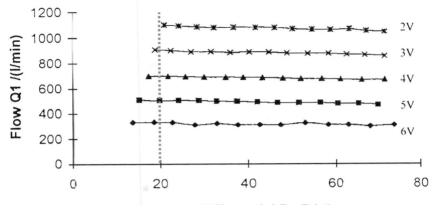

Figure 2.119. Motor drive modified steady-state performance

The system steady-state motor speed characteristic for the selected linearisation conditions are shown in Figure 2.119.

Note that the motor speed is now only slightly affected by load pressure. The complete steady-state efficiency picture is shown in Figure 2.120.

Unfortunately this much improved steady-state characteristic hides the fact that the linearisation process has removed almost all system dynamic damping since the contribution from the flow control valve has been almost eliminated by design. However, since a pressure transducer is need to continually monitor the line pressure it may be used to create digital filtering in software. In this case it can be shown that a high-pass filter will stabilise the system and can be designed to give a suitable transient behaviour. The controller is also used to low-pass filter the

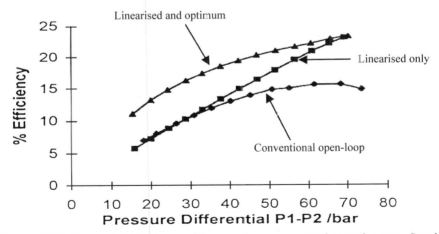

Figure 2.120. Steady-state efficiency of the open-loop motor speed control system. Supply pressure = 100 bar, motor speed 500 rpm

supply pressure transducer signal to minimise noise. Hence the two pressures are filtered as follows:

Supply pressure filter transfer function $\dfrac{1}{\left(1+s\tau_1\right)}$ (2.269)

Line pressure filter transfer function $\dfrac{s\tau_2}{\left(1+s\tau_2\right)}$ (2.270)

These transfer functions are transformed within the PSC automatically into digital filter algorithms and for this application a sampling interval of 4 ms was used to produce the best performance. The transient response to a step change in speed demand over the range 500 rpm to 800 rpm is shown in Figure 2.121 for the original system and for the modified system with intelligent control.

A feel for system dynamics may be obtained by considering the ideal case with perfect servovalve flow nonlinearity cancellation and no motor damping. The effect of fluid compressibility, volume V on either side, and motor load inertia J

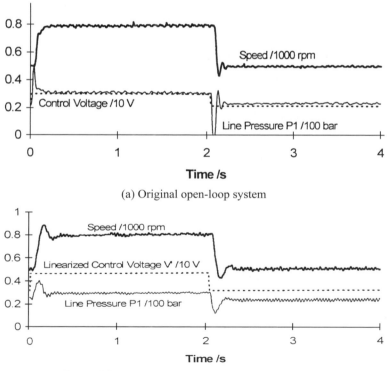

(a) Original open-loop system

(b) Modified open-loop system with intelligent control

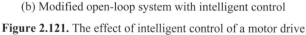

Figure 2.121. The effect of intelligent control of a motor drive

then results in the following transfer function relating motor speed to servovalve applied voltage:

$$\frac{\omega}{V_e} = \frac{\dfrac{K}{D_m}}{1 + \dfrac{JVs^2}{2\beta D_m^2}} \qquad K = \frac{k_f Ga}{D_m}\sqrt{\frac{P_s}{2}} \tag{2.271}$$

Considering the pressure loop and the filter transfer function (2.270) then results in the following modified open-loop transfer function:

$$\frac{\omega(s)}{K V_e(s)} = \frac{(1+s\tau)}{\left[1+s\tau+s^2\left(\dfrac{1}{\omega_n^2}+\dfrac{KJH_p\tau}{D_m}\right)+s^3\dfrac{\tau}{\omega_n^2}\right]} \tag{2.272}$$

$$\omega_n = \sqrt{\frac{2\beta D_m^2}{JV}}$$

The gain of each pressure transducer is H_p. The introduction of dynamic pressure differential damping, via an electronic filter, has ostensibly resulted in a stable open-loop response but increased the order of the transfer function from 2 to 3. This suggests that the ITAE criterion can probably be used to set the system variables K and τ. To check stability of (2.272) the usual Routh array method may be applied. It will be deduced that the modified transfer function is inherently stable and independent of τ. Hence a suitable value of τ may be selected to give the best dynamic performance. However in practice the intelligent control process briefly described does result in a low bandwidth mainly due to noise and motor ripple effects, dynamics of the servovalve, and pressure control valve dynamics not considered here.

Chapter 3

Condition Monitoring Methods

3.1 Methods Based Around Steady-state Flow Loss Metering

3.1.1 Visual Leak Detection, the Simplest Approach

Visual inspection of leaks is perhaps the simplest approach to condition monitoring and at a relatively low cost. The Spectroline leak detection system for oil-based fluids, for example, is a highly successful approach:

- it uses a 20 ml fluorescent additive [Oil-Glo^{TM}] per 5 litre oil
- the additive is UV light sensitive
- it enables leaks to be detected using a portable 100 w inspection lamp

Figure 3.1 shows a visual result of seal leakage detection from a cylinder using this method.

Figure 3.1. The Oil-Glo^{TM} method of leakage detection [Advanced Engineering Ltd Basingstoke UK]

Once added, it will remain in the system and only 3 grams/litre fluid is required which makes it extremely economical in use. Checking for leaks clearly requires just a matter of seconds using a robust yet lightweight hand-held lamp.

3.1.2 Flow Meter Types

Flow, pressure and vibration monitoring are perhaps the dominant techniques for condition monitoring of fluid power systems. Sensors are placed at strategic points in a circuit, the type of sensor being selected to suit the application. For steady-state flow measurement the flow meter could be a simple low-cost variable area type with a moving visual indicator, yet still able to operate at high pressures. However, for *in situ* applications the flow meter is more likely to be of the gear-type, turbine-type or poppet sensing type. These tend to be more accurate, have electronic sensing elements, have on-board signal processing features, digital read-out and the facility for

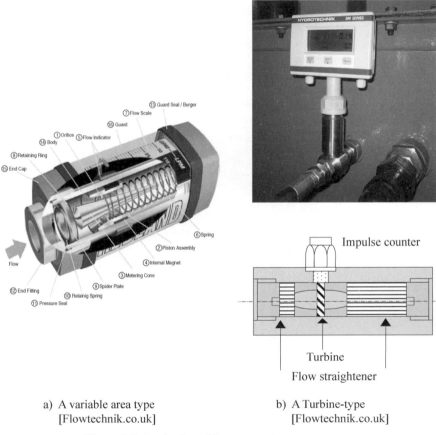

a) A variable area type
[Flowtechnik.co.uk]

b) A Turbine-type
[Flowtechnik.co.uk]

Figure 3.2. A selection of flow meters in common use

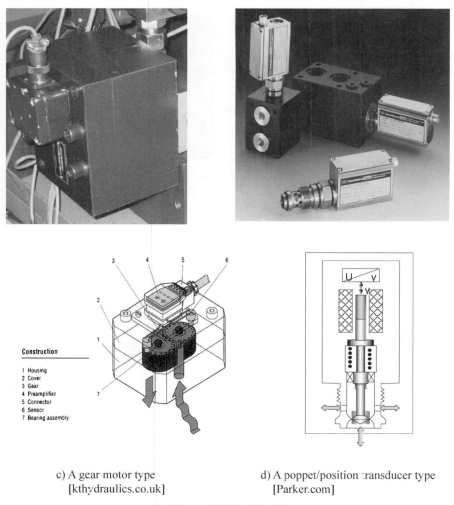

Construction

1 Housing
2 Cover
3 Gear
4 Preamplifier
5 Connector
6 Sensor
7 Bearing assembly

c) A gear motor type
[kthydraulics.co.uk]

d) A poppet/position transducer type
[Parker.com]

Figure 3.2. (continued)

interfacing with a computer or hand-held data acquisition unit. A selection of the four common types are shown in Figure 3.2(a)–(d).

The range of flow metering is now quite staggering, for example 0.005 L/min at the low end of the spectrum, gear type, to 18200 L/min at the high end of the spectrum, turbine type.

3.1.3 Vane Pump Wear Monitoring

Consider an example arising from a serious problem in a steel rod-drawing application. A study on the continually-failing vane pumps, operating with a water-based

Figure 3.3. Vane pump stator damage when running with a water-based fluid

fluid, revealed cavitation problems at low temperatures in the plant. Although the pump power levels are small at 7.5 kW, the resulting plant downtime created unacceptable costs. The solution was simply to heat the fluid reservoir during expected cold weather periods. Laboratory tests were undertaken on new and worn pumps and Figure 3.3 shows one such pump in its badly worn condition and running with a 5/95 oil/water emulsion.

A steady-state flow/pressure test would reveal the damage as shown in Figure 3.4 which compares data after 40 and 71 running hours.

In addition, surface roughness measurements were taken on the stator and weight loss measurements were made for the stator, vanes and rotor. Some results are shown in Figure 3.5. As expected the weight loss data validates the flow data in the sense of what is causing the flow change, although in this application the use of flow metering would be inappropriate and a better option would be vibration measurement as will be discussed later.

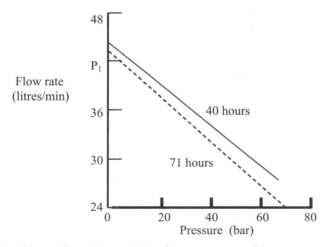

Figure 3.4. Steady-state flow characteristic of a vane pump under severe wear conditions, 9/95 emulsion

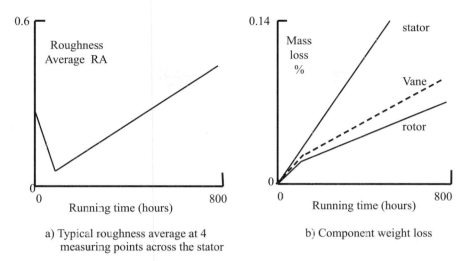

a) Typical roughness average at 4
measuring points across the stator

b) Component weight loss

Figure 3.5. Measurements of wear for a vane pump running on 5/95 emulsion

3.1.4 Assessing the External Leakage from a Pressure-compensated Axial Piston Pump

This pressure-compensated pump is operating with a 3/97 emulsion in a de-rated condition up to a maximum of 160 bar for a steel mill continuous casting application. It is well known that pumps operating with high water content fluids have a poor efficiency, as is noted from the previous example for a vane pump.

In this study a turbine type flow meter was used to measure flow rate. A torque meter was available and connected between the electric motor drive shaft and the pump drive shaft. Pressure transducers were also placed at the inlet and outlet lines. The pump assembly is shown in Figure 3.6.

Figure 3.6. Measuring the flow characteristic of an axial piston pump

It is known that:

- leakage losses are proportional to pressure difference
- the centrifugal boost pump provides a constant inlet pressure of 3.5 bar
- the pressure compensator internal pressure is constant as governed by the fixed swash plate setting for this test
- the test temperature varied between 35°C and 38°C

The experimental procedure was as follows:

- Vary the load pressure and measure the input torque, external leakage flow rate, output flow rate (two flow meters required, external leakage flow Figure (3.2b), output flow Figure 3.6).
- Determine the pump displacement D_p by noting in this case a negligible drop in motor drive speed (a 37 kW electric motor was used, 1440 rpm).
- The pump steady-state theory outlined in Chapter 2 allows calculation of the pump external loss resistance R_e from the total external leakage measurement.
- Hence from the output flow measurement the internal, cross-port, leakage resistance R_i may be determined.

The measured torque and flow rate characteristics are shown in Figure 3.7.

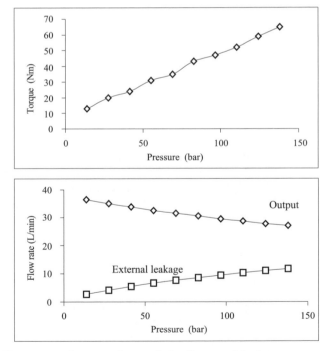

Figure 3.7. Flow rate and torque characteristics for an axial piston pump operating with a 3/97 emulsion

The calculations resulted in the following data:

- a pump displacement, calculated from the torque measurements, of $D_p = 4.19 \times 10^{-6}\,m^3/rad$. This compares with that measured from geometry of $D_p = 4.05 \times 10^{-6}\,m^3/rad$
- external leakage resistance $R_e = 14\,bar/L\,min^{-1}$
- internal (cross port) leakage resistance $R_i = 280\,bar/L\,min^{-1}$
- a fixed pressure compensator flow leakage of $2.5\,L\,min^{-1}$
- a maximum pump efficiency of 65.6% at a pressure of 95 bar. Using 35 cSt mineral oil would give a maximum efficiency of 89% at a pressure of 242 bar.

For the practical application at a working pressure of 140 bar the following conclusions are reached:

- the efficiency is 60%
- the pump external leakage flow rate is $10\,L/min$
- the pump cross port leakage flow rate is $0.5\,L/min$
- the compensator flow rate leakage is $2.1\,L/min$

3.1.5 Detection of Cylinder Piston Seal Wear for a Servovalve/Cylinder Drive

In this example the circuit is tested in its unloaded condition. The leakage flow rate across the seal may be deduced from measurements of the inlet and outlet flow rates. Using the notation in Figure 3.8 and the background theory developed in Section 2.7 gives:

Extending

$$P_2 - P_1 = \frac{P_s\,(\gamma - 1)}{\left(1 + \gamma^3\right)} \tag{3.1}$$

$$\text{Seal leakage} = \frac{A_2 Q_1 - A_1 Q_2}{A_{rod}}$$

Retracting

$$P_2 - P_1 = \frac{P_s \gamma^2\,(\gamma - 1)}{\left(1 + \gamma^3\right)} \tag{3.2}$$

$$\text{Seal leakage} = \frac{A_1 Q_2 - A_2 Q_1}{A_{rod}}$$

It can be seen that in the unloaded condition the pressure differential acts from the rod side to the bore side for both extending and retracting cases. In fact for this study since the area ratio $\gamma = 1.46$, the pressure differential when retracting is just over double that when extending. Seal leakage calculations based on measurement

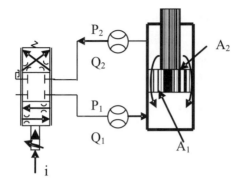

Figure 3.8. Servovalve/cylinder drive

are therefore expected to be more reliable for the retracting condition, particularly at small levels of wear. Figure 3.9 shows the result of a comparison between a new seal and a worn seal in an unacceptable condition.

An external restrictor is also shown in Figure 3.9, the reason being to show what values of seal leakage are needed to stop retraction of the cylinder in the unloaded condition. It will also be observed that very low computed leakages are evident with a new seal in place. This application also reveals the importance of comparing the seal performance at the same operating conditions. It also indicates a small actuator velocity effect, as evident from the increasing leakage at increasing

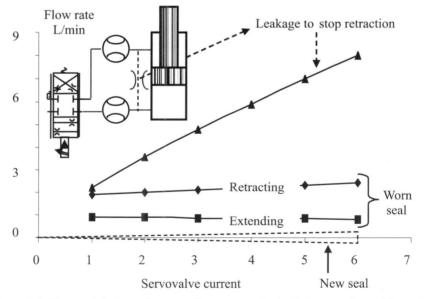

Figure 3.9. Measured leakage flow rate for a servovalve/cylinder open-loop drive using inlet and outlet flow metering, $\gamma = 1.46$

servovalve currents, and this will be considered in a later example using dynamic flow measuring techniques.

3.1.6 Assessing Motor Leakage Characteristics by Coupling to a Servovalve

In a similar circuit to that in the previous example, next consider a motor as actuator. By measuring the inlet and outlet flow characteristics on-line, any deterioration in cross port and external leakage may be determined. Chapter 2, Section 2 developed the steady-state flow rate equations and some measured flow rate characteristics. The test circuit is shown as Figure 3.10 and the sum of line pressures is shown here as Figure 3.11.

Recalling the steady state flow rate equations given in Chapter 2

$$Q_1 = D_m\omega + \frac{(P_1 - P_2)}{R_i} + \frac{P_1}{R_e}$$

$$Q_2 = D_m\omega + \frac{(P_1 - P_2)}{R_i} - \frac{P_2}{R_e}$$

(3.3)

It follows from the data and the, simplified, theory that the following deductions seem reasonable for this study for the speed and pressure ranges appropriate to the test conditions:

$$\text{Flow difference } Q_1 - Q_2 = \frac{(P_1 + P_2)}{R_e} \cong \frac{0.97\, P_s}{R_e}$$

$$\cong 0.5\, \text{L/min}$$

(3.4)

$$\text{Mean flow } \frac{Q_1 + Q_2}{2} = D_m\omega + \left(\frac{1}{R_i} + \frac{1}{2R_e}\right)(P_1 - P_2)$$

(3.5)

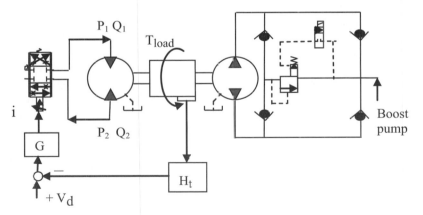

Figure 3.10. Servovalve/motor with loading circuit

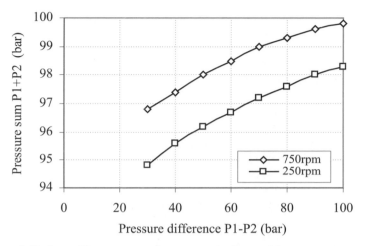

Figure 3.11. Sum of line pressures for a servovalve/motor drive at constant speeds

So, the external resistance may be deduced from (3.4) allowing the internal resistance, and the motor displacement, to be deduced from (3.5) once the mean flow line is constructed on the flow characteristic. For this study, and for a motor in its virtually new condition, the measured data are as follows:

$$R_e \approx 1.5 \times 10^{12}\, \mathrm{Nm^{-2}/m^3 s^{-1}}$$

$$R_i \approx 3.3 \times 10^{12}\, \mathrm{Nm^{-2}/m^3 s^{-1}}$$

$$D_m = 1.59 \times 10^{-6}\, \mathrm{m^3/radian}.$$

These parameters may be checked on-line to see if leakage paths are changing due to wear or other internal damaging mechanisms.

3.2 Cylinder Seal Leakage Identification Within a Vehicle Active Suspension Actuator Using Dynamic Data

This study considers a ¼ car active suspension unit with a servovalve/double-rod cylinder as the active actuator and supplied by a racing car manufacturer. It is assembled within a test rig that combines a commercial car wheel and suspension linkage as shown in Figure 3.12. The control system incorporates two computers, one for the road input for the active actuator and incorporating a MoogM3000 digital Programmable Servo Controller (PSC) card. This allows both complex control strategies to be implemented and also nonlinear/logic characteristics to be generated in real-time. The second computer is used to monitor performance using a National Instruments data acquisition card and associated LabView software. A schematic of the instrumentation approach is shown in Figure 3.13.

Figure 3.12. A ¼ car active suspension test rig. Cardiff University

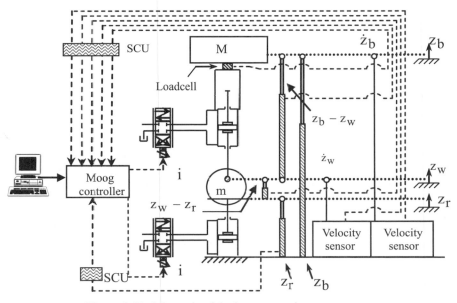

Figure 3.13. Schematic of the instrumentation arrangement

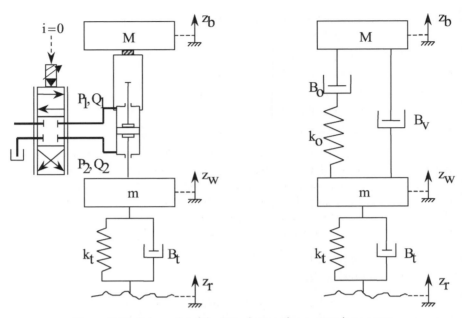

Figure 3.14. Schematic of the open-loop active suspension system

The supplied active actuator is in a well-used condition although its exact condition is unknown. It also contains an integral rod position transducer. Additional instrumentation is added to measure the position and velocity of the body, wheel, road together with a load cell to measure the load force. Where appropriate, signals may be conditioned via the Signal Conditioning Unit (SCU), for example to filter high-frequency noise.

The vehicle body is constrained to move in a vertical direction via linear bearings. Both the active actuator and the road input position actuator are servovalve controlled, the latter within a conventional position control loop. A schematic of the open-loop system is shown in Figure 3.14.

Following a linearised analysis of this system, the active actuator oil compressibility and leakage terms act in series as shown in Figure 3.14 together with a parallel viscous friction term. Low-cost "Synflex" flexible lines were used between the servovalve and active actuator and detailed dynamic analysis tests on very long lines, using transmission line theory, revealed an expected low value of effective bulk modulus.

The parameters and additional data are as follows:

- ¼ car mass $M = 240$ kg, wheel mass $m = 40$ kg
- tyre stiffness k_t measured as 2.8×10^5 N/m
- tyre damping coefficient B_t measured dynamic tests as 4000 N/ms^{-1}
- actuator viscous damping coefficient B_v, to be determined from this study

- actuator seal/servovalve cross-line leakage coefficient $B_o = A^2 R_i$ – to be determined from this study. R_i is the leakage resistance.
- the actuator is inclined at an angle $\alpha = 27°$ to the vertical
- effective bulk modulus $\beta = 0.22 \times 10^9 \, N/m^2$
- actuator area $A = 2.46 \times 10^{-4} \, m^2$, actuator/hose volume $V = 7.13 \times 10^{-5} \, m^3$
- actuator oil/line equivalent hydraulic stiffness $k_o = 2\beta A^2 / V = 3.73 \, N/m$
- supply pressure $P_s = 200 \, bar$, load pressure $P_{load} = 108 \, bar$

The approach here is to use linearised theory to develop the open-loop transfer function and then to use an identification technique linked to the theory and transient response practical tests foe wheel position changes, that is the active actuator only is controlled in a closed-loop mode. Now it is known from Chapter 2 that the flow gain changes with direction of motion due to either the assistance or resistance of the body force. It will be recalled that the ratio of flow gains is given by:

$$\frac{\text{Retracting flow gain}}{\text{Extending flow gain}} = \sqrt{\frac{P_s + P_{load}}{P_s - P_{load}}} = 1.83 \tag{3.6}$$

The Moog PSC can easily be programmed to insert a directional gain to pre-multiply the signal to the servovalve. In this application the extending gain was increased by a factor of 1.35 and the retracting gain was decreased by a factor of 1.35. Consider then the closed-loop digital control system shown in Figure 3.15.

This one degree of freedom, in a vibration mode sense, then results in a closed-loop transfer function as follows:

$$\frac{z_b}{z_b \, \text{demand}} = \frac{\varepsilon_o}{\left(a_o + a_1 s + a_2 s^2 + a_3 s^3\right)}$$

$$a_o = 2AF_1F_2Gk_iNR\cos\alpha \quad a_1 = \frac{2B_v}{R_i} + 2A^2\cos\alpha \tag{3.7}$$

$$a_2 = \frac{2M}{R_i} + \frac{B_v V}{\beta} \qquad a_3 = \frac{MV}{\beta}$$

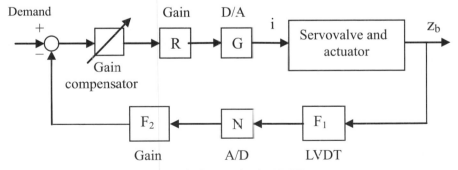

Figure 3.15. Block diagram for the 1DOF system test

Further PSC data are as follows:

- $F_1 = 57.2\,\text{V/m}$, $F_2 = 1$
- $N = 1600\,\text{PSCno/V}$, $G = 6.24 \times 10^{-3}\,\text{mA/PSC no}$
- $R = 0.4$
- The servovalve linearised flow gain $k_i = 2.3 \times 10^{-5}\,\text{m}^3\text{s}^{-1}/\text{mA}$

A road input position demand step was applied and the displacement at the wheel used as the model input. The MATLAB® identification toolbox was used to identify the transfer function coefficients using practical time response data and the prediction error method. This resulted in:

$$a_o = 2.30 \times 10^{-6} \quad a_1 = 1.14 \times 10^{-7} \quad a_2 = 5.00 \times 10^{-9} \quad a_3 = 7.78 \times 10^{-11} \quad (3.8)$$

Typical measured and identified transient responses are shown in Figure 3.16. It follows from (3.7) and (3.8) that the identified leakage resistance and viscous damping coefficient are:

Viscous damping coefficient $B_v = 300\,\text{N/ms}^{-1}$

Leakage resistance $R_i = 9.8 \times 10^{10}\,\text{Nm}^{-2}/\text{m}^3\text{s}^{-1}$ $\qquad\qquad\qquad\qquad$ (3.9)

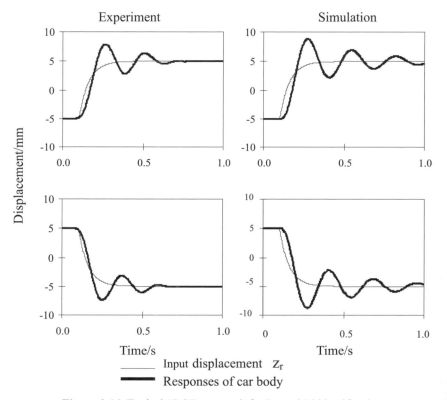

Figure 3.16. Typical 1DOF test result for Bv and Ri identification

A two degree of freedom test with active control switched-in produces a much more complex sixth-order transfer function relating body position to actual road position. This allows validation of the tyre stiffness k_t, which can easily be measured, and the tyre damping coefficient B_t.

3.3 Fluid Borne (FBN), Structural Borne (SBN) and Air Borne (ABN) noise

3.3.1 Introduction

The very operation of fluid power systems gives rise to FBN, SBN, ABN noise and their relative importance depends upon the specific type of component and/or fault condition. Reducing noise levels, particularly for pumps and motors, is a continuous activity by manufacturers as noise regulations become more rigorous for working environments. Human hearing can recover from short periods of high noise levels but it is known that continuous exposure will result in a permanent hearing loss. A general guidance is that audible noise should be reduced below a level of 85 dB, and sound pressure levels above 200 Pa must be avoided (Peters, 2002). Clearly fluid power component noise generation as the source of the problem must be given a high priority at the development stage, and continual development is put into this area by manufacturers. Consideration of both fluid mechanics and mechanical design is helped by a similar development in modelling techniques.

Noise will be generated in hydraulic systems due to:

- Fluid turbulence
- Cavitation due to sudden drops in pressure followed by an increase in pressure
- Inherent ripple such as in the flow generation characteristic of a pump
- Badly designed or ill-considered dynamic performance of systems creating pressure instability giving rise to another ripple component
- Fault conditions generating internal mechanical damage and possibly a deteriorating dynamic behaviour such as wear of a poppet valve

The noise generated can be one of four main types as shown in Figure 3.17.

A variety of sensors are available to measure FBN and SBN levels and measurements, other than ABN, are usually converted to a deciBel (dB) scale defined simply as:

$$dB = 20\log\left[\frac{u}{u_{ref}}\right] \tag{3.10}$$

where u is the measurement and u_{ref} is the reference level. The appropriate reference levels are given in Table 3.1. Note that some instruments may use 10 rather than 20 in equation (3.10).

Noise type	Typical causes

Periodic

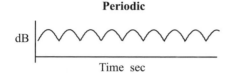

Pump and motor pressure ripple, fluid vortex shedding, flow induced vibration, vibration of loads, out of balance effects

Steady random

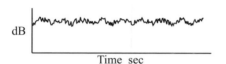

Fluid turbulence, separation, high velocity gradients, rotational background noise, room background noise

Time varying

Cavitation, effect of components being switched on/off, acoustic emission due to crack development, nonlinear control behaviour

Periodic impulsive

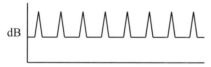

Faulty components such as vanes, pistons, bearings, gears

Figure 3.17. Origins of noise in fluid power systems

Table 3.1. Noise reference levels for various measurement parameters

Measurement parameter	Reference level
displacement	10^{-11} m
velocity	5×10^{-8} m/s
acceleration	10^{-5} m/s^2
general voltage	1 V (rms)
acoustic pressure	20×10^{-6} Pa

3.3.2 Application to Pumps, FBN and SBN Spectrum Analysis

Measurement and frequency spectrum analysis of pump pressure would appear to be a convenient method for the possibility of fault detection. However, a pump is just part of a system and in practice care has to be taken in interpreting the data. Consider, for example, an axial piston pump as briefly described in Section 2.3. The ideal flow source is a rectified sine wave superimposed onto the mean flow rate as shown in Figure 3.18.

The peak-to-peak value of piston-generated flow ripple, δQ, is given by:

$$\delta Q = \frac{\pi}{n} \tan \frac{\pi}{2n} \qquad \text{n even}$$

$$\delta Q = \frac{\pi}{2n} \tan \frac{\pi}{4n} \qquad \text{n odd}$$

(3.11)

where n is the number of pistons (Thoma, 1964). For typical values of $n = 7$ and 9, the ripple magnitude is 2.5% and 1.5% of the mean flow rate. Note the advantage of having an odd number of pistons when considering this ripple as shown in Figure 3.19.

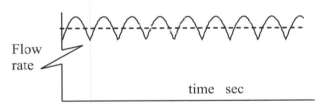

Figure 3.18. Ideal flow source for an axial piston pump

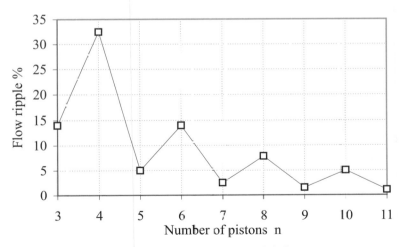

Figure 3.19. Ripple magnitude for an axial piston pump

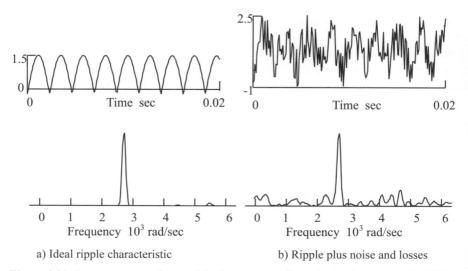

Figure 3.20. Power spectra of an axial piston pump flow ripple. Pump speed = 24 Hz, n = 9 pistons, pumping frequency = 1357 rad/sec

This ideal flow source, obtained by summing individual flow rate contribution for each piston, produces a frequency of nN_p where n is the number of pistons and N_p is the pump speed. This value of nN_p is often called the pumping frequency. However, the signal is rectified so that a frequency spectrum of this signal identifies a frequency of $2\,nN_p$. Figure 3.20 shows this rectified flow source without and with noise. Also shown are the frequency spectra for both cases, and for a pump speed of $Np = 24$ Hz with $n = 9$ pistons. In both cases one spectral component at twice the pumping frequency can be seen, even in the presence of severe noise levels.

In practice the frequency spectrum measurement has to be carefully considered in relation to the sensor being used. This data relates to flow which is difficult to directly measure over the required frequency range for very large flow rates. If pressure sensing is used then the pump and connected system must be considered if the pressure spectrum is to be sensibly explained. The pressure ripple generated by the ideal flow source is as a result of the dynamic interaction between source, relief valve and load. A detailed analysis and understanding requires the concept of fluid impedance to be introduced and this will be discussed later.

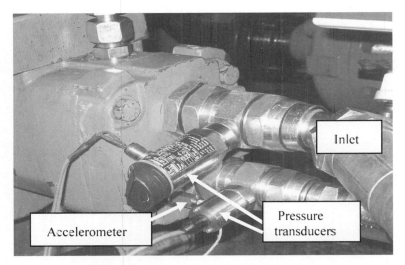

Figure 3.21. Monitoring and axial piston pump

An alternative approach to fault monitoring is to use an accelerometer or acoustic emission sensor attached to the pump body. In this case some of the body modes of vibration may well be indicated in the frequency spectrum in addition to any fault indicators. The spectrum content is also dependent on the type of pump being assessed, its size and the manner in which it is mechanically coupled to the system. For example, Figure 3.21 shows an accelerometer and a pressure transducers attached to an axial piston pump operating with a 3/97 emulsion.

Figure 3.22 shows accelerometer and pressure spectra for the pump at a speed of 1493 rpm (24.9 Hz) and with nine pistons. The pumping frequency is 224 Hz.

The results are obtained via a National Instruments data acquisition system using LabView software and a frequency zoom of 450±30 Hz is shown. It can be seen from Figure 3.22 that the expected frequency of $2\,nN_p = 448\,Hz$ is measured and found to be dominant using both an accelerometer and a pressure transducer. Note also that the pressure transducer is detecting a further system frequency component at around 450 Hz that need to be identified.

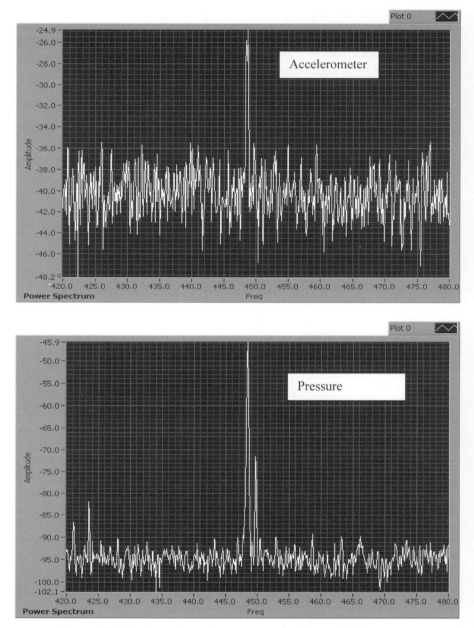

Figure 3.22. Accelerometer and pressure spectra for an axial piston pump

3.3.3 Fault Data Trending

A common method of following a potential fault development is to track a measurement, for example the frequency component magnitude such as that shown in Figure 3.22. The change in magnitude is then recorded at suitable intervals and the trend analysed using methods now made easy by the use of modern signal processing packages and computer graphics. Consider a particular measurement of pump pressure peak-to-peak noise fluctuation using day averages taken from the new condition, and then with a suspected developing wear problem. Typical data are shown in Figure 3.23, each sample being averaged from several samples during the day in practice. An alternative measurement could be the signal root-mean-square (rms) value.

Using run-in data over a 10 day period, and from day 4, the ripple peak-to-peak standard deviation $\sigma = 0.24$ bar and the ripple mean value $\mu = 9.0$ bar. It can be seen that there appears to be a possible small change occurring after 20 days, but alternative ways of looking at the data must be used to statistically validate this suggestion. Three common methods will now be considered.

(i) Moving average, using $n = 3$ days for each average in this example
It is common to use $\pm 3\sigma$ limits to indicate alarm levels and taking into account the sample size, the levels are given by:

$$\text{alarm levels} \quad \mu \pm \frac{3\sigma}{\sqrt{n}} = 9.0 \pm 0.42\,\text{bar} \tag{3.12}$$

These limits are shown in Figure 3.24 where a fault alarm is indicated after 23 days of operation.

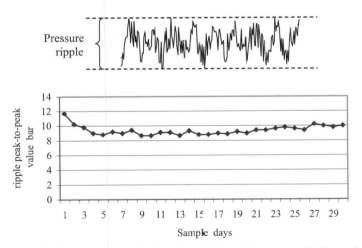

Figure 3.23. Pump pressure ripple and its daily variation over a 30 day period

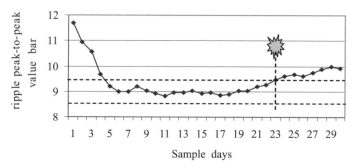

Figure 3.24. 3-day moving average for pump pressure ripple

(ii) Exponentially weighted mean average (EWMA)

$$z(t) = \lambda x(t) + (\lambda - 1)z(t - T) \tag{3.13}$$

where $x(t)$ is the measurement magnitude, $z(t)$ the computed mean using the current sample, $z(t-T)$ the computed mean at the previous sample, T is the data sampling interval and $0 < \lambda < 1$. In this study $\lambda = 0.2$. The alarm levels are usually set to the following moving values:

$$\text{Alarm levels} \qquad \mu \pm 3\sigma \sqrt{\frac{\lambda \left[1 - (1 - \lambda)^{2i} \right]}{(2 - \lambda)}} \qquad i \text{ is the sample number}$$

$$\text{The asymptote is} \quad \mu \pm 3\sigma \sqrt{\frac{\lambda}{(2 - \lambda)}} \quad \text{as } i \rightarrow \infty \tag{3.14}$$

For this study calculations are shown in Figure 3.25. The asymptotes only are shown since they are rapidly approached for a large sample size and hence away from the run-in condition. Again, an alarm is indicated after 23 days of operation.

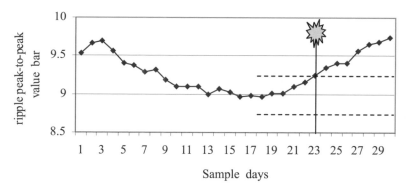

Figure 3.25. Exponentially weighted moving average representation of the pump pressure ripple data

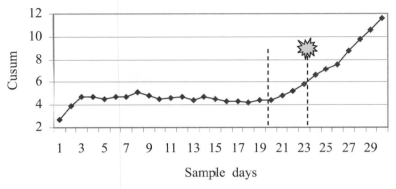

Figure 3.26. Cusum for pump pressure ripple

(iii) Cumulative Sum (Cusum), where the change from an assumed mean is simply added to the value calculated for the previous sample. In this example the assumed mean $\mu = 9.0$ bar. A developing fault is expected to produce a consistent bias in a measurement and therefore the Cusum would be expected to highlight this continuing bias. Figure 3.26 shows the result for this study.

The Cusum approach suggests a developing problem after 20 days, reinforcing the conclusion reached using the 3-day moving average approach. Bearing in mind the 3-day moving average approach and the EWMA approach, it would seem from Figure 3.26 that an increase in Cusum by a factor of 1.5 gives a good indication of the alarm condition.

This example illustrates perhaps the need to consider more than one approach to fault level indication, although it does seem from other studies that the Cusum and EWMA methods are the preferred methods.

3.3.4 Acoustic Emission Sensor Application and Comparison for Pump Wear and Cavitation Detection, SBN and FBN

Acoustic emission sensors can often give information in applications where vibration or pressure sensors appear to be relatively inadequate. They can also be small in size, such as that shown in comparison to other high-quality sensors in Figure 3.27.

High frequency stress wave signals are transmitted through the connecting material and their characteristics change as either the vibration stimulation characteristic changes or the material properties change at the microscopic level. These signals are detected by a piezoelectric element in conjunction with integral, specially designed high-frequency preamplifiers. They are therefore selected with a frequency sensing characteristic that best matches the fault phenomenon to be measured. Consider then stator wear detection on a vane pump as previously discussed in Section 3.1.3, and operating with a water-based emulsion. Figure 3.28 shows an acoustic emission sensor instantaneous signal for the pump in its new

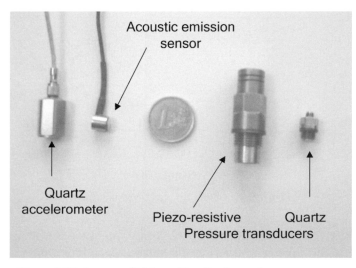

Figure 3.27. A range of high-quality vibration and pressure sensors

condition and a badly worn condition. The pump speed is 960 rpm and since there are 10 vanes the pumping frequency is 160 Hz. An on-line frequency spectrum computation *using an accelerometer* showed just harmonics of pumping frequency with apparently no suitable fault indicator frequency. However, observing the higher frequency part of the spectrum revealed potentially useful information which must be assumed to be related to a body mode of vibration, and initially

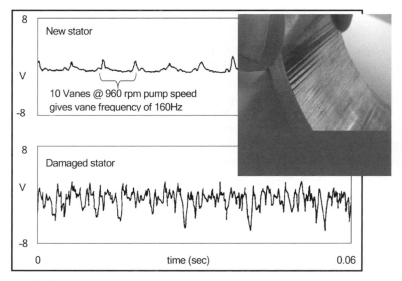

Figure 3.28. Time domain signals from an acoustic emission sensor used to detect vane pump stator wear

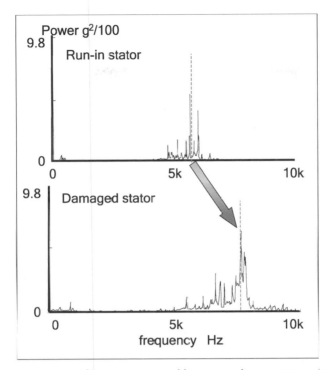

Figure 3.29. Power spectra for a vane pump with a new and worn stator using an accelerometer

centred around a frequency of approximately 6 kHz. This centre frequency increases with stator wear as shown in Figure 3.29.

Considering now the acoustic sensor mean voltage output rather than instantaneous results in a very efficient fault indicator as shown in Figure 3.30.

The large variation in sensor output, at any desired reference pressure, shows that smaller rates of deterioration detection are possible using this approach.

For completeness, a pressure transducer signal will now be considered. Figure 3.31 shows the pump ripple with the obvious distortion visually evident for the worn stator. This visual approach is not appropriate for use as an on-line method and therefore it is again worth pursuing a frequency spectrum approach.

Before this is illustrated, it is appropriate to consider the stator damage shown in Figure 3.3 and 3.28. The ridging effect is dominant at the two flow inlet sections of the stator but does not have a definitive characteristic dimension. A measurement estimate suggests a value of typically 1 mm giving 200 ridges if they appeared completely around the periphery of the stator. Hence this could be interpreted as an additional pressure frequency component having a value of $200 \times 16 = 3200$ Hz. This is not similar to the centre frequency indicated by the

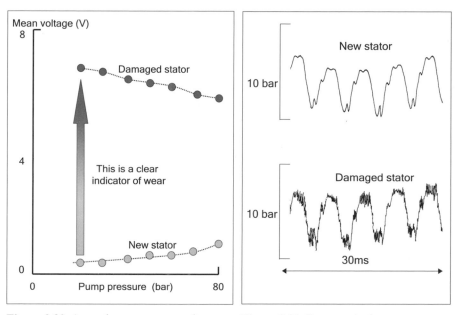

Figure 3.30. Acoustic sensor mean voltage output

Figure 3.31. Pressure ripple

accelerometer measurement shown in Figure 3.29 for either the pump in its new or worn condition. This centre frequency moves from a frequency of 6000 Hz in the new condition to a higher value of 7700 Hz for the badly worn condition. Consider then the pressure frequency spectrum shown in Figure 3.32 at a working pressure of 70 bar.

It can be seen from Figure 3.32 that little information on wear can be deduced at the pumping frequency of 160 Hz. However, it would appear that the "ridge frequency" of around 3200 Hz is being detected by the pressure sensor, and can clearly be tracked with wear change. It does therefore seem that for this study the acoustic sensor mean voltage output is a very useful indicator of wear. Otherwise pressure transducer spectra are useful but must be analysed at whatever ridge frequency is generated by the wear.

A further indication of the power of an acoustic sensor output mean voltage can be deduced from Figure 3.33 which shows the measurement from an acoustic emission sensor placed at the inlet of a pump. The pump inlet flow is deliberately, but briefly, restricted to create undesirable cavitation conditions. In these tests the pump normal inlet pressure of +0.1 bar is briefly reduced to –0.2 bar and the acoustic emission sensor output mean voltage is shown in Figure 3.33 for a pump outlet pressure of 70 bar with mineral oil operating at a temperature of 70°C. The performance of the sensor is self-evident, and significantly better than the use of an accelerometer and its subsequent signal processing.

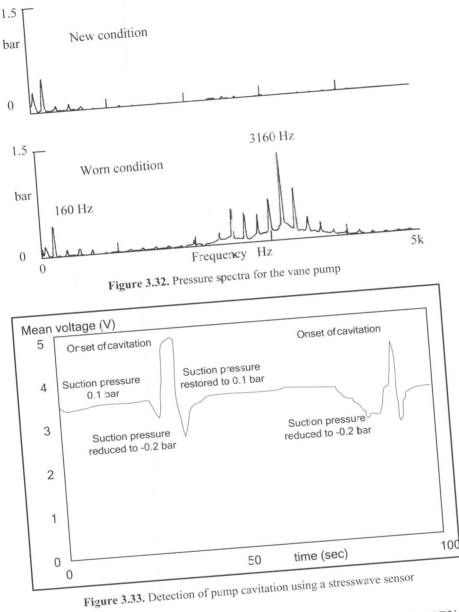

Figure 3.32. Pressure spectra for the vane pump

Figure 3.33. Detection of pump cavitation using a stresswave sensor

3.3.5 Spectrum Analysis to Detect Piston Seal Wear within a Cylinder, SBN

Now consider the use of an accelerometer fixed to the body of a single-rod cylinder
controlled by a solenoid-operated four-way directional valve. The previous examples used power spectrum plotted on a linear scale, but in this cylinder study the

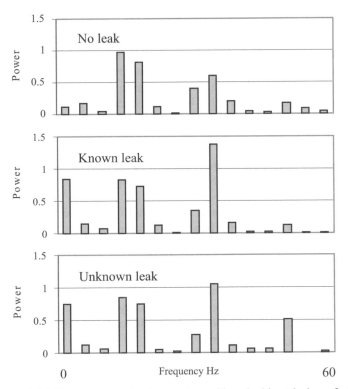

Figure 3.36. Frequency spectra for a system with and without leakage faults

15 lines for this first analysis. All three spectra have common frequency components and it is not easy to say to which archetype the unknown leakage condition is linked. One statistical way of doing this is to compute a correlation function (CF) between two sets of n spectral lines defined as follows:

$$CF = \frac{\left[\sum\limits_{i=1}^{n} ab\right]^2}{\left[\sum\limits_{i=1}^{n} a^2\right]\left[\sum\limits_{i=1}^{n} b^b\right]} \qquad (3.15)$$

This CF value will fall within the range $0 < CF < 1$ and it is possible to determine the fault origin by seeking the highest value for any spectra pairing. Table 3.2 shows the CF matrix and also for more detailed data with 60 spectral lines.

In both cases the strongest correlation is to the known leakage fault condition rather than the no-fault condition. It is deduced that the new measurement spectrum is indicative of a leakage fault condition. This approach however, as with all diagnostic techniques, becomes more complex for the multiple-fault case.

Table 3.2. CF values for three spectra

15 lines	No leakage	Known leakage	unknown leakage
No leakage	1.00	0.72	0.77
Known leakage	0.72	1.00	**0.93**
Unknown leakage	0.77	**0.93**	1.00
60 lines	No leakage	Known leakage	Unknown leakage
No leakage	1.00	0.51	0.47
Known leakage	0.51	1.00	**0.93**
Unknown leakage	0.47	**0.93**	1.00

3.3.7 Component Impedance, FBN Analysis

When components are linked together the overall impedance concept is a powerful small signal analysis method, particularly if pressure spectra are used for diagnostics. The impedance approach allows the determination of the dominant dynamic characteristic and can also be used to identify an unknown impedance where direct measurement is impossible, such as the source impedance of a pump. Consider a basic pump/lumped volume/pressure relief valve/load as shown in Figure 3.37.

The equivalent impedance model is shown where the source flow Q_p is the ideal flow generation characteristic, Z_s is the source characteristic of the pump, Z_{rv} is the impedance of the pressure relief valve, and Z_{load} is the impedance of the load. From flow rate continuity, and including fluid compressibility:

$$Q_p = Q_s + Q_{rv} + Q_{load} + C\frac{dP}{dt}$$
(3.16)

where C is the fluid compressibility equivalent capacitance. Since impedance is defined as:

$$\text{Impedance } Z = \frac{P}{Q}$$
(3.17)

It then follows that the overall impedance Z is given by:

$$\frac{1}{Z} = \frac{1}{Z_s} + \frac{1}{Z_{rv}} - \frac{1}{Z_{load}} + \frac{1}{Z_c}$$
(3.18)

The problem in practice is the determination of each impedance, particularly for components where dynamic effects become evident at high frequencies and where noise analysis becomes important. Direct measurement in the frequency domain using fast acting flow meters has already been discussed in Chapter 2 and earlier

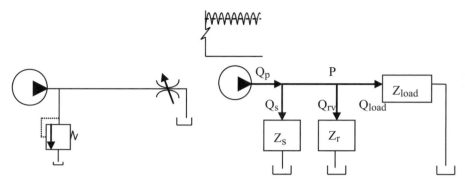

Figure 3.37. Impedance representation of hydraulic components

in this section. For pump source impedance the consideration of transmission line theory allows this evaluation using a number of transducers strategically placed along the line. Figure 3.38 shows some formative results for an axial piston pump, a gear pump, and a flow control valve. Considering a pump, it would be expected that the impedance characteristic should have perhaps four regimes as the frequency increases and conceptually described as follows:

- Low frequency, the impedance is purely resistive due to leakage:

    $Z_s \rightarrow R$,   the pump leakage resistance

- Frequency increasing, the impedance becomes influenced by fluid compressibility effects:

    $Z_s \rightarrow \dfrac{1}{j\omega C}$ where C is a defining capacitance

- At higher frequencies, the impedance becomes influenced by fluid inertia effects:

    $Z_s \rightarrow j\omega L$ where L is a defining inductance

- At very high frequencies, the impedance is influenced by transmission line-type effects

These regimes are evident for both the pumps and also the flow control valve in Figure 3.38. Over the frequency range considered, the phase changes initially from $-90°$ to $+90°$ (only frequencies above 100 Hz are shown) indicating a change from compressibility effects to inductive effects. At frequencies beyond 2.5 kHz the computations become difficult to impossible due to the experimental technique but more significantly due to the inherent poor quality of the measured data. However, there is clear evidence of more complex dynamic effects, possibly transmission line-type in the small chambers.

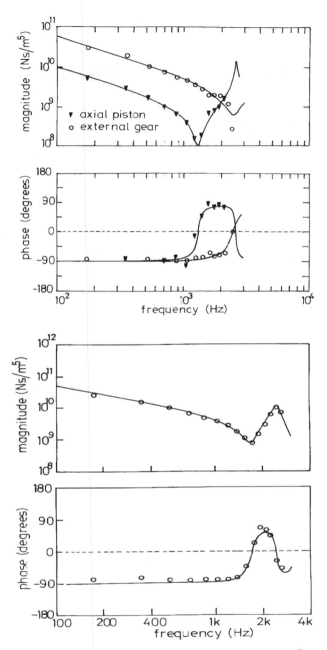

Figure 3.38. Source impedance of an axial piston, external gear pump, flow control valve [Edge and Johnston 1986]

3.3.8 Use of Impedance for Fault Diagnosis of a Pressure Relief Valve Using Frequency Response

Recalling modelling techniques discussed in Chapter 2, a time-series data-based approach may be used to determine the impedance transfer function. The data-based approach has been discussed in some detail in Section 2.18 where it was shown that the pressure relief valve linearised impedance could be accurately determined. Figure 3.39 shows a schematic of the pressure relief valve and a particular fault associated with malfunctioning of the crucial damping element.

Given that the valve is continually loading and unloading, the input dynamic pressure and flow rate may be used to continually update the dynamic impedance. The earlier analysis showed that the impedance may be written in the following sampled data form:

$$\frac{\delta P(z)}{\delta Q(z)} = \frac{\left(b_o + b_1 z^{-1} + b_2 z^{-2}\right)}{\left(1 - a_1 z^{-1} - a_2 z^{-2}\right)} \tag{3.19}$$

The damping unit is now considered in two stages of deterioration, a minor fault where the damper unit is partially open (due to small particles in practice), and a major fault where the damper unit is essentially stuck open. Figure 3.40 shows the comparison of on-line determined impedance with the valve in its correctly-operating condition. Changes in both magnitude and phase may be deduced, perhaps peak values and associated frequency being useful as an indicator in practice. The actual coefficients in (3.19) are shown in Table 3.3.

It is deduced from Table 3.3 that the denominator coefficients (operating on pressure) barely change whereas the numerator coefficients (operating on flow rate) significantly change with fault severity. This information could be used in conjunction with the frequency response information to develop a knowledge-based approach to diagnosis.

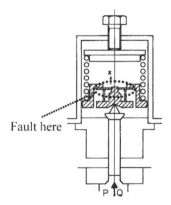

Fault here

Figure 3.39. Schematic of the pressure relief valve

Table 3.3. Pressure relief valve impedance discrete transfer function coefficients for increasing fault severity

Coefficient	Good condition	Minor fault	Major fault
a_1	3.12	3.25	3.14
a_2	−3.23	−3.49	−3.26
a_3	1.12	1.24	1.13
$b_0/10^9$	−10.3	−7.63	−1.6
$b_1/10^9$	29.7	21.4	3.94
$b_2/10^9$	−28.6	−20.1	−3.14
$b_3/10^9$	9.2	6.3	0.8

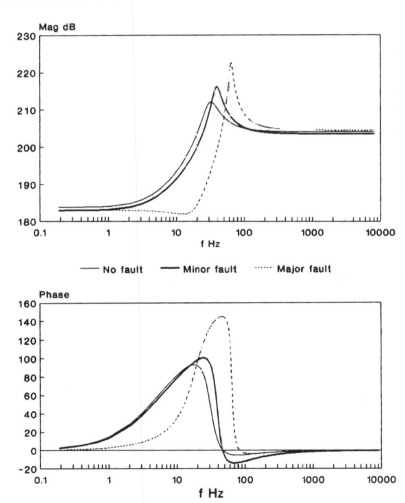

Figure 3.40. On-line computation of a pressure relief valve input impedance as a potential fault indicating method – damping unit fault

3.3.9 SBN due to Repetitive Fault Phenomena

When monitoring hydraulic systems, particularly on manufacturing plant, it is inevitable that information will also be gathered from rotating components. It therefore makes sense to also consider condition monitoring of such components since it is a simple matter to include the information into the existing knowledge base. SBN analysis has already been discussed in the context of signal analysis using accelerometers and acoustic emission sensors. Considering a shaft speed N rpm, cyclic phenomena in practice may be caused by:

- out of balance   N
- misalignment or bent shaft   N, probably 2 N, could be 4 N
- damaged gear   N
- looseness   2 N
- oil film whirl $\approx 0.46\,N$
- ball race defects

Consider a ball race type of bearing, Figure 3.41 with a pitch circle diameter D, ball diameter d, and ball contact angle β. For such a bearing mechanical damage can occur on any of the moving parts, which in the general case can be inner race, outer race, ball, cage. All are turning at different frequencies but related to the drive shaft frequency assuming normal surface contact behaviour. Assuming a shaft drive speed N and the number of ball elements n, then it can be shown that each frequency is given by:

Outer race defect frequency $\dfrac{nN}{2}\left[1-\dfrac{d}{D}\cos\beta\right]$ (3.20)

Inner race defect frequency $\dfrac{nN}{2}\left[1+\dfrac{d}{D}\cos\beta\right]$ (3.21)

Ball defect frequency $\qquad N\dfrac{D}{d}\left[1-\left(\dfrac{d}{D}\cos\beta\right)^{2}\right]$ (3.22)

Cage defect frequency $\qquad \dfrac{N}{2}\left[1-\dfrac{d}{D}\cos\beta\right]$ (3.23)

Consider an application concerned with on-line monitoring of a no-twist rod finishing mill, the final transmission to each roll pair being via 10 precision bevel gearboxes. This work followed from earlier work on vane pump monitoring described earlier in this chapter and resulted in a much more advanced on-line implementation (Morgan and Watton, 1990). A vibration analysis of such a system needs to take into account all possible mechanical failure possibilities, which in this case may be caused by gears, ball race bearings, rolling elements and transmission shafts. A close collaboration with the mill manufacturer is essential to determine all data necessary

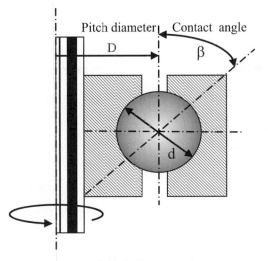

Figure 3.41. Ball race bearing

for determining faults. For example, all frequency modes for the large transmission shafts are required and in practice this is achieved via Finite Element Analysis.

For this particular application all component operating frequencies are normalised with respect to the drive frequency and it just turns out that each frequency ratio has a unique value. This is helped by the fact that as the steel rod passes through the mill, the speed of each pair of rolls must be increased to achieve mass flow continuity. Hence each gearbox has different combinations of teeth. A photograph of the mill with the guard up, and hence not in operation, is shown in Figure 3.42 and a schematic of the bevel gearbox drive arrangement is shown as Figure 3.43.

Accelerometers were attached to each gearbox housing as shown in Figure 3.44 and hard-wired to the central process computer via a frequency analyser in combination with a signal multiplexer. The vibration data is continually monitored and an expert system approach used to determine the most probable cause of a defect vibration should one exist.

The drive speeds are between 800 rpm and 1200 rpm and the normalised frequency calculations are only accepted if the speed measurement before and after the calculations are within ±2 rpm. The final system considered 400 different frequency components, which were coded to allow quick identification of the appropriate part of the mill. A simple computer screen graphics was developed to indicate the fault in conjunction with a green, amber, red alert colour code. Vibration velocity alarm levels were set initially as follows:

peak < 2 mm/s   Green

2 < peak < 4   Amber

4 < peak   Red

Figure 3.42. No-twist rod finishing mill [formerly ASW Ltd Cardiff UK]

One particular fault recorded is shown in Figure 3.45.

The fault was examined during the next maintenance period, the fault recognised as shown in Figure 3.45 and the bearing replaced. The subsequent frequency spectrum showed no evidence of the previous frequency component. This new approach allowed a change in maintenance practice such that a priority of fault checking was established rather than the previous practice of automatically

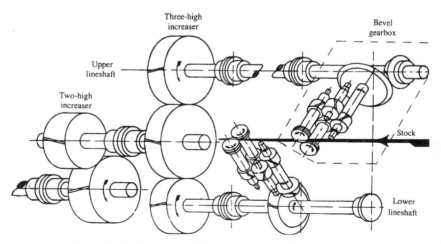

Figure 3.43. Schematic of the drive system of shafts and gears

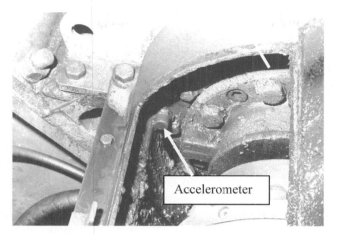

Figure 3.44. Accelerometer placement on a bevel gearbox housing

changing bearings on a rotational basis irrespective of whether or not a fault existed. This project was financed on the company assumption of a 2 year payback period. Due to fault detection of the type shown earlier and the resulting reduction in downtime and its associated costs overviewed in Chapter 1, *the actual payback period was 5 months*. The last reported saving over a 5 year period was a factor of 100 greater than the investment cost. Interestingly the concept of a maintenance overhead has appropriately changed to a maintenance investment and two similar systems were then installed at another rod mill.

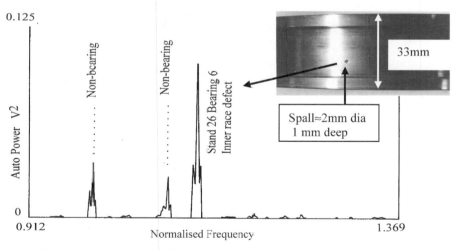

Figure 3.45. Inner race defect, stand 26 bearing 6 [Morgan and Watton, 1990]

3.3.10 ABN Concepts

The noise emitted by a hydraulic component is received by the human ear and the final test of good noise design is the sound pressure level and its perception within the operating environment. When considering a sound level meter recording, the human ear frequency perception must be taken into account and this is done via a frequency weighting, usually the A weighting spectrum, which is programmed into the meter. A modern sound level meter is shown in Figure 3.46 together with the A weighting scale.

Noise measurements are made using a microphone coupled to the analyser and are usually computed using either an octave band, 1/1, analysis or a 1/3 octave band analysis. This is different to the narrowband analysis of signals previously discussed when analysing FBN or SBN. The use of 1/12 octave analysis is sometimes employed with some signal analysers to more adequately represent the ideal spectrum. The analyser shown in Figure 3.46, for example, employs a 1/1 analysis centre frequency 8 Hz to 16 kHz and a 1/3 octave analysis centre frequency 6.3 Hz to 20 kHz and has a dynamic range of more than 135 dB. It is an advanced instrument embodying many features required for comprehensive noise analysis, and capable of real-time measurements at intervals from 1 second to 24 hours. When considering a noise measurement, the background level must be taken into account. It must be remembered that calculations are made on a logarithmic scale if the true noise level is required or if the cumulative effect of adding two components with known noise levels is required. The sound power level is defined as:

$$L = 10\log\left(\frac{W}{W_{ref}}\right) \qquad W_{ref} = 10^{-12}\,\text{watts} \tag{3.24}$$

Note that if sound pressure level is to be used then equation (3.10) usually applies. The reference pressure used for ABN analysis corresponds to the minimum audible sound at 1 kHz, an absolute pressure of 20 μPa, and is designated the 0 dB

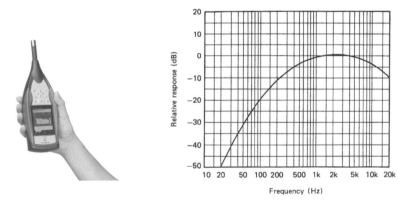

Figure 3.46. Bruel & Kjaer Type 2250 hand-held analyser and the A weighted scale

sound pressure level. The human ear is able to detect a sound pressure level up to 200 Pa and this factor of 10^7 gives rise to a decibel range of 140 dB, the threshold of pain. This of course applies to measured data filtered using the A weighting scale.

Noise measurements are sensitive to the position and orientation of the sound level meter in addition to the background noise level. Measurements are usually taken at a number of points around a hypothetical sphere of radius 1 m, if possible, and the data averaged. Quarter sphere and half sphere measurements have also been suggested as adequate for testing in an anechoic chamber with up to eight readings being sufficient. For measurements within an industrial environment it is crucial that a measurement strategy is agreed by all concerned and that measurements are taken from consistent positions. Figure 3.47 shows 1/3 octave measurements for a quieter restrictor valve and compares a new design, using a variable annular tapered flow passage, with the original design, using a conventional needle valve.

The dramatic improvement in the noise behaviour can be seen, the overall noise level being 39.6 dB(A) compared with 69.7 dB(A) for the original needle valve.

A comparison of some mean power levels for different pumps is shown in Figure 3.48 and for a range of operating pressures. These results are useful in clarifying the typical, and perhaps expected, variations in noise level as the operating power capacity of a pump is increased as reflected by the pump design type.

High operating power levels, particularly pressure, usually require the use of axial piston pumps and the inherent design results in higher noise levels when

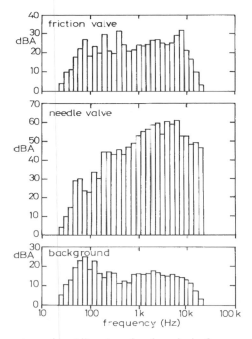

Figure 3.47. ABN spectra using 1/3 octave band analysis for a restrictor valve [Silva, 1991]

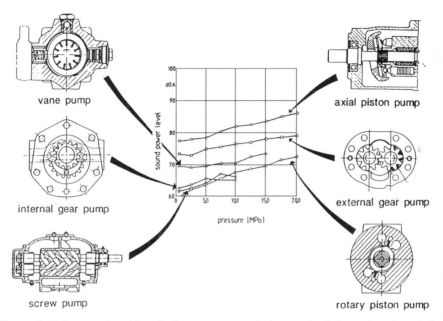

Figure 3.48. Range of positive displacement pump designs and relative noise characteristics [Bergmann, 1990]

compared to other relatively simpler designs. It can be seen that incorrect pump selection could result in a 15 dB increase in noise level, assuming the pressure and flow rate may be achieved for a variety of pump types.

To determine the noise level of an operating component then the total noise level L_t and the background noise level L_b, with the component not in operation, may be used to determine the unknown component noise level L_c. Using the addition of power levels, the component noise level above the background level is given by:

$$L_c - L_b = 10\log\left[10^{\frac{(L_t - L_b)}{10}} - 1 \right] \tag{3.25}$$

Equation (3.25) is plotted as Figure 3.49.

It can be seen that this correction is only required when the difference $L_t - L_b$ is less than typically 10 dB. Otherwise the logarithmic addition effect is negligible.

This approach to component fault monitoring using ABN information is perhaps difficult to implement due to the practical issue of noise data gathering and also the variable background noise. However, some form of regular average noise level measurement is now seen as essential from an environmental health point of view.

The European Community Directive requires the protection of all persons at work with actions required depending on the noise level. Noise assessments must be made with records being kept and continually updated by competent persons.

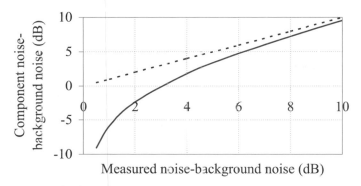

Figure 3.49. Determination of true component noise knowing the noise levels with and without the component in operation

Adequate information, instruction and training should be given regarding the steps to be taken to minimise the risks from exposure to high noise levels. Hearing protectors must be maintained and repaired as required. Hearing protection is required, particularly for daily exposure levels between 85 dB(A) and 90 dB(A) with special precautions to be taken for impulsive noise situations creating pressure levels of 200 Pa or more. Protection should also include the availability of emergency exits and rescue arrangements from the place of work as well as fire prevention and smoking control arrangements inside and around the space.

3.4 Time Encoded Signal Processing, TESP Analysis

3.4.1 Introduction

The TESP approach utilises time-based data and in essence extracts dynamic features of the signal via a novel signal processing technique. By using a pre-selected and consistent amount of data in the time domain, dynamic characteristics of the signal may be deduced and compared in the presence of a fault. Dynamic features extraction is best illustrated by example. Consider therefore a signal composed of three frequency components as follows:

$$s(t) = \cos t - 0.01 \cos 2t - 0.31 \cos 3t \qquad (3.26)$$

Using the substitution:

$$\cos \omega t = \frac{e^{j\omega t} + e^{-j\omega t}}{2} \text{ and let } x = e^{j\omega t} \qquad (3.27)$$

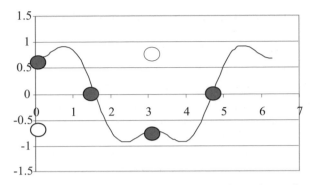

Figure 3.50. Time signal $s(t) = \cos t - 0.01 \cos 2t - 0.31 \cos 3t$ and complex time roots of $s(x) = 0$

Now find the roots of this equation, that is:

$$s(x) = \frac{1}{2x^3}(-0.31x^6 - 0.01x^5 + x^4 + 0x^3 + x^2 - 0.01x - 0.31) = 0 \qquad (3.28)$$

Now define an important concept, that of **complex time:**

$$t = \tau + j\sigma \qquad (3.29)$$

and hence $x = e^{j\omega t} = e^{-\omega\sigma}(\cos \omega\tau + j \sin \omega\tau) \qquad (3.30)$

The roots of (3.28) and the corresponding complex time mappings are:

roots x	mapped complex time $\tau + j\sigma$
1.979	−0.683j
0.505	**0.683j**
−2.008	**3.142 − 0.697j**
−0.498	3.142 + 0.697j
−0.0052 + j	**1.576**
−0.0052 − j	**4.707**

By choosing a suitable scaling factor for both the original time signal and the complex time mappings, they may then be overlayed and compared. It will be noticed that four of the six values (shown in bold above) match the time signal as seen in Figure 3.50.

The following mappings may be deduced:

time signal →	s(t)	complex time roots
$t = 0$	$s(t) = 0.68$	$t_2 = 0 + 0.68j$
$t = 1.58$	$s(t) = 0$	$t_5 = 1.58 + 0j$
$t = 3.14$	$s(t) = -0.7$	$t_3 = 3.14 - 0.7j$
$t = 4.71$	$s(t) = 0$	$t_6 = 4.71 + 0j$

This leads to the next important illustration that **within the zero crossings** of the time signal, **the turning point**(s) is/are also mapped to a complex time root. The outcome of this explanation by example is:

- If a time-varying signal is considered, zero-base it, then the zero crossings and minima within them are a measure of the dynamic properties of the signal.
- If the signal dynamic properties change due to a fault, then the hypothesis is that these characteristic shape parameters also change and it should be possible to use this to classify the fault(s).
- The time between zero crossings is called an **epoch**, and therefore the TESP approach in its simplest form is based upon just **2 shape parameters**.
- The diagnostic art is to know what to do with these shape parameters (**D** the time duration, and **T** the number of turning points within the epoch) as they change with fault conditions.

3.4.2 Example. Pump Torque Data Analysis

Consider the computer-sampled torque data from a pump. Pump input torque may be measured by using an in-line torque transducer mounted via a pair of flexible couplings and the most common transducers are strain gauge type, optical type, and surface acoustic wave type.

Surface acoustic wave (SAW) torque sensors are relatively new and have the advantage that that they are a non-contact type. The transducer shaft has two interdigital arrays of thin metal electrodes deposited on a highly polished piezoelectric substrate. A radio frequency signal applied at the input effectively creates surface waves (Rayleigh waves) and changes in resonant frequency of the electrodes due to torque-induced strain is used to determine the torque (Sensor Technology Ltd UK). Figure 3.51 shows a rotary torque transducer and its positioning

Figure 3.51. Rotary torque transducer [Sensor Technology Ltd UK] and placement in a pump drive

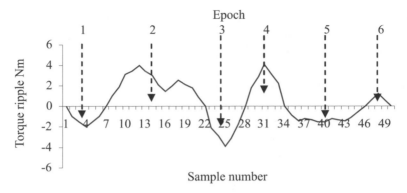

Figure 3.52. Torque fluctuation for a run-in vane pump

within an axial piston pump drive. Note that when using such a method the torsional characteristic of the sensor and couplings must be taken into account to ensure that coupling modes of vibration are not influencing decisions on faults that are not attributable to them.

Consider then the torque data with a computer sampling frequency of 1 kHz and a time record of 50 samples as shown in Figure 3.52. The actual signal must be zero-mean based to apply TESP analysis and often this can be visually estimated. In this case a mean torque of 10 Nm has been removed for the no-load condition currently being analysed. The actual mean torque measured was 10.31 Nm but the error in this case is not important in the calculations. For on-line applications the pre-processing of data is easily automated.

For each epoch the Duration (**D**) and Turning points (**T**) may be determined and these are as follows:

		Duration **D** (ms)	Turning points **T** (number)
epoch 1		7	0
epoch 2		15	1
epoch 3		6	0
epoch 4		6	0
epoch 5		12	2
epoch 6		4	0
mean value	μ	8.33	0.5
standard deviation	σ	3.86	0.76

The condition (or fault) **Archetype** must now be statistically normalised to ensure that sensible comparisons are made between data groups. In practice the following approach has been found to capture the significant information:

- consider $\mu \pm 2\sigma$ of data
- divide the data into 10 classes

Considering the Duration data
The data range is 8.33 ± 7.72 $0.61 \rightarrow 16.05$
This is then divided into 10 classes with a class width of $4\sigma/10 = 15.44/10 = 1.54$.

Considering the Turning point data
The data range is 0.5 ± 1.52 $-1.02 \rightarrow 2.02$
This is then divided into 10 classes with a class width of $4\sigma/10 = 3.04/10 = 0.3$.

Both **D** and **T** data are transformed as shown in Figure 3.53.

To classify a fault, the data for the pump operating with a suspected fault must next be considered and then compared with the appropriate archetype or archetypes. In this example it is done by comparing **D** and **T** data using a simple correlation function (CF) as described in the previous section. The worn pump torque ripple is shown in Figure 3.54 and the classified **D** and **T** data are shown in Figure 3.55.

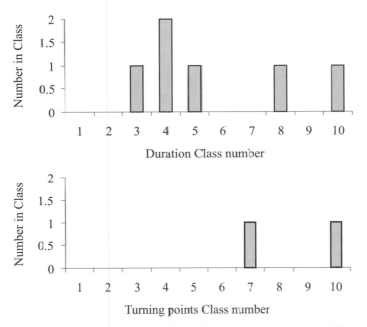

Figure 3.53. Pump torque classified archetypes for the run-in condition

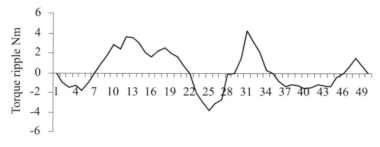

Figure 3.54. Worn pump torque ripple about the mean value

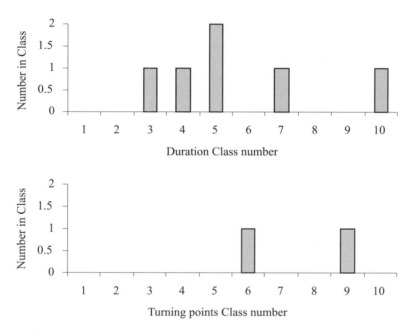

Figure 3.55. Pump torque classification for a suspected wear condition

We might first consider comparing basic data such as mean and standard deviation as follows:

Duration **D**		Turning points **T**	
New condition	Fault condition	New condition	Fault condition
mean = 8.33	mean = 8.33	mean = 0.5	mean = 0.83
$\sigma = 3.86$	$\sigma = 3.64$	$\sigma = 0.76$	$\sigma = 0.9$

It can be seen that there is little to be gained from this information, particularly from Duration data, although the Turning points mean has increased. However, this may not seem a good way forward for classification purposes. Therefore consider the correlation function the TESP classification data and considering both new and fault conditions gives the following result:

For **Duration** $CF(new, fault) = \dfrac{36}{8 \times 8} = 0.56$

For **Turning** points $CF(new, fault) = \dfrac{0}{1 \times 5} = 0$

So actually **D** is middle value, **T** is zero but sparse of data, and the combination show probably very little correlation. It is deduced that the perceived fault has probably just started to develop and must be carefully monitored. Clearly from a diagnostics point of view this basic concept must be extended to include further dynamic information with the introduction of alarm level conditions to indicate the move from correlation to no-correlation as a fault develops. This will now be pursued.

3.4.3 Example. Application to 28 Servovalves on the Work Roll Bending Control System of a Seven-stand Hot Steel Strip Mill

A schematic of a typical rolling stand is shown in Figure 3.56 and illustrates the three main hydraulic control systems.

This application is concerned with just the WRB hydraulic pressure control systems, each of the seven stands having four servovalves per stand. Typical pressure and servovalve signal transients are shown if Figure 3.57 as the steel strip enters and leaves a stand, and a TESP analysis has been undertaken using such data from all seven stands and for many hours of operation.

The servovalve wear is known and due to flapper/nozzle erosion. Hence the early detection of this wear will avoid expensive plant downtime and prioritise servovalve repair/replacement work. One approach developed in parallel with this work is simply to detect the valve instability frequency via Fast Fourier Transform analysis of the signals. In fact, this has been implemented in real time, as discussed further in Chapter 4 on Expert Systems, and has been extremely successful. It is therefore useful here to compare the TESP approach with the more conventional approach using frequency spectrum analysis tracking the fault frequency.

Note from Figure 3.57 the rapid increase in WRB pressure as the strip enters the stand with pressure control ensuring an ostensibly constant value while rolling. The servovalve control voltage rapidly drops to around zero under pressure control and the spikes upon entering and leaving the stand may be seen. Note also that the transient pressures look identical for the new servovalve and the worn servovalve, although a noisier signal can be seen for the worn servovalve voltage. Additional work on the servovalve via laboratory tests has shown that the servovalve voltage signals tend to give better fault classification results than the pressure signals,

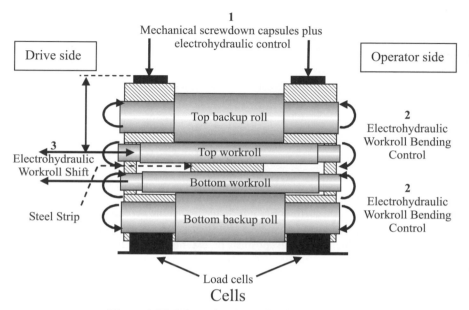

Figure 3.56. Schematic of a steel strip rolling stand

though both work well. The TESP classification approach previously described is now modified to take into account additional signal characteristics and as follows:

- **D** duration of each epoch
- **A** peak amplitude within the epoch
- **T** the number of turning points within the epoch
- **I** absolute area within the epoch

The CRj between each measurement and the new condition archetype is evaluated for each j = **D, A, T, I** and the standard deviation, σ_j is evaluated for each new condition archetype. A fault detection coefficient, FDC, is then computed as follows:

$$\text{FDC} = \sum_j \frac{\text{CR}_j \, \sigma_j}{\sum_j \sigma_j} \quad j = \text{D, A, T, I} \tag{3.31}$$

Mill computer alarms are then set at the following values:

☐ Green satisfactory $\text{FDC} \geq 0.75$

◻ Amber, warning $0.5 \leq \text{FDC} \leq 0.75$ $\tag{3.32}$

◼ Red, alert $\text{FDC} < 0.5$

Several methods of data averaging and comparison have been used, all actually giving very similar results. The simplest approach of comparing a current measurement with the previous measurement has been found to be remarkably

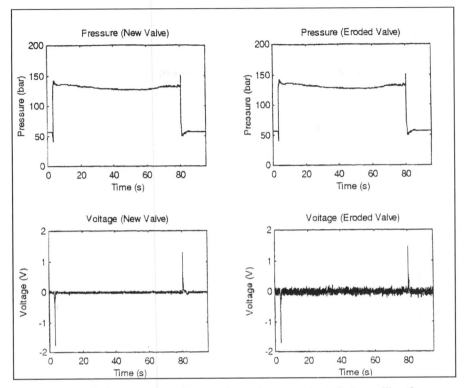

Figure 3.57. Cylinder pressure and servovalve voltage transients during rolling for a new servovalve and a worn servovalve

good resulting in essentially a moving archetype approach. For this application the plant provided data at a sampling frequency of 50 Hz and no other signal processing was undertaken.

A frequency spectrum approach is somewhat limited with this data although spectrum changes within the 25 Hz range are evident for a deteriorating servovalve. Therefore 25 classes were selected for the frequency spectrum approach and typical data are shown in Figure 3.58. For the TESP application 12 classes were used and some typical measured DATI archetypes and measurements are shown in Figure 3.59.

The TESP approach does seem to give better signal discrimination then the FFT approach although it would appear that the FFT approach could still be a useful tool. Consider then the 28 servovalves monitored every hour and displayed over a 31 hour period as shown in Figure 3.60. The notation for each servovalve is as follows:

- OSB Operator Side Bottom
- OST Operator Side Top
- DST Drive Side Bottom
- DSB Drive Side Top

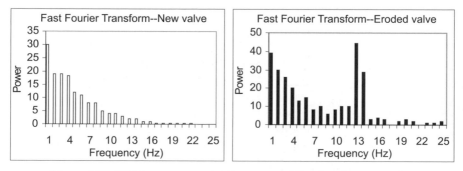

Figure 3.58. FFT frequency spectra for a new and deteriorating servovalve

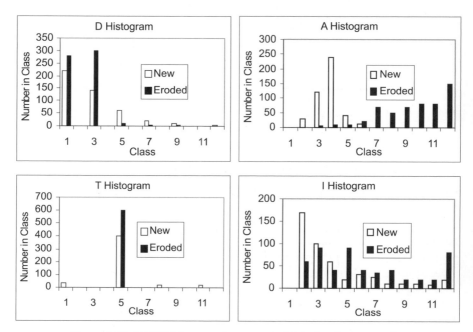

Figure 3.59. DATI histograms for a new and deteriorating servovalve

The FFT approach has identified a probable dominant wear effect for Stand 5 DSB, but it can clearly be seen from the TESP approach that this servovalve together with four other servovalves are actually exhibiting a deterioration in performance.

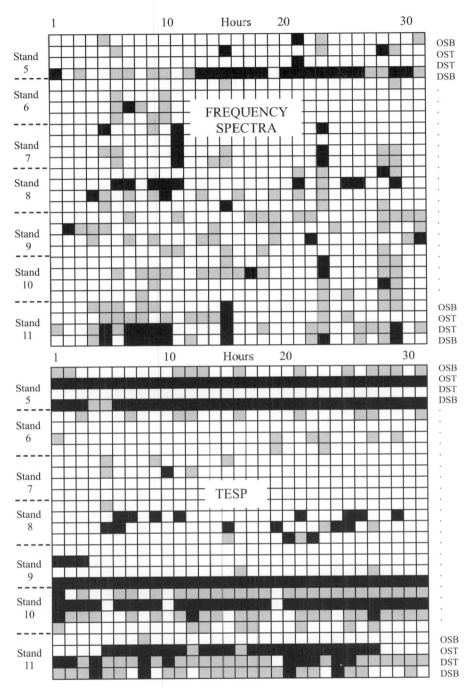

Figure 3.60. Frequency spectrum and TESP approach for servovalve fault detection on a 28-servovalve rolling mill control system

3.4.4 Combined TESP/ANN Approach to Leakage and Servovalve Fault Detection in a Pressure Control System

Consider the pressure control system that was previously analysed in Section 3.3.6 which was concerned with a frequency spectrum correlation technique and for line leakage only. The diagram is reproduced as Figure 3.61.

The system has a faulty servovalve possibility, in fact the same as analysed from the rolling mill study discussed in the previous section, together with a fractured line possibility. Tests were carried out in a laboratory environment and with servovalves with and without erosion faults taken from the previously discussed rolling mill environment. Although there only two fault possibilities, this multiple fault possibility is often a diagnostic challenge for fluid power systems. The principle of analysis here is:

- Apply TESP to the pressure transient and the servovalve drive voltage transient. In this case the pressure signal is numerically differentiated and the servovalve error signal has any bias voltage removed.
- Generate DATI data for pressure and servovalve voltage, that is, 40 classes for pressure and 40 classes for servovalve voltage.
- Select a suitable ANN and apply the 80 classes, now called Car-code, as parallel inputs.
- Classify different faults using known fault training data.

The concepts behind the use of ANNs for modelling and fault diagnosis were discussed in Section 2.20. Considering the approach here, the ANN topology schematic is shown in Figure 3.62.

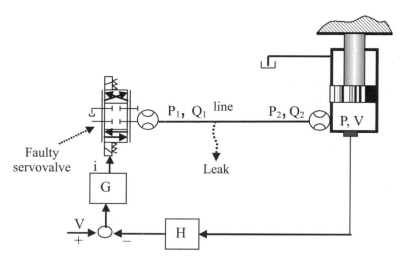

Figure 3.61. Pressure control system

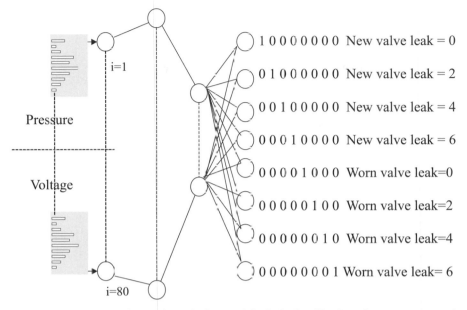

Figure 3.62. ANN topology schematic for multiple-fault classification of a pressure control system with a possible leak and faulty servovalve, and using Car-code via TESP analysis

It will be seen from Figure 3.62 that a servovalve was used with and without damage and each also in combination with leakage flows of 0, 2, 4, 6 litres/min. It is often more cautious to develop a separate ANN for each fault and this of course reduces the number of outputs. This has been done in this study together with a single ANN to embrace all faults as indicated in Figure 3.62. In all cases it was found necessary, following many trials. to use two hidden layers and many further trials were necessary to optimise the number of neurons in each hidden layer. Only 500 iterations were found to be necessary for training and to avoid over-training unseen data were used every 20 iterations. The best topologies were found to be as follows:

- Leakage faults only   80–20–4–4   98% identification success
- Servovalve faults only   80–20–1–2   100% identification success
- Combined faults   80–20–8–8   100% identification success

Each ANN performance was obtained using 30 new data sets. Typical data and Car-codes for a good servovalve are shown in Figures 3.63 and 3.64.

The ANN performance using unseen data is shown in figures 3.65 and 3.66.

This approach, once optimised as shown, gives excellent discrimination between fault conditions, even at the lowest leakage levels used of 2 litres/min.

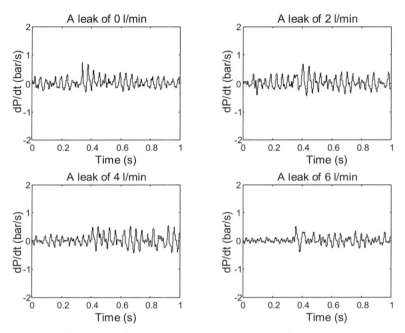

Figure 3.63. Pre-processed signals, undamaged servovalve

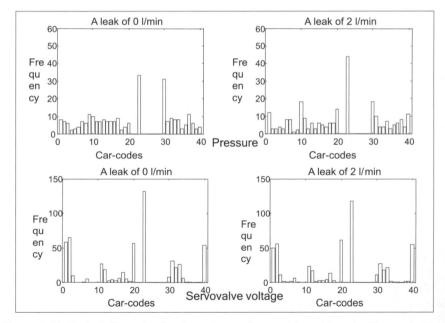

Figure 3.64. Typical Car-codes for the undamaged servovalve with and without a leakage of 2 litres/min

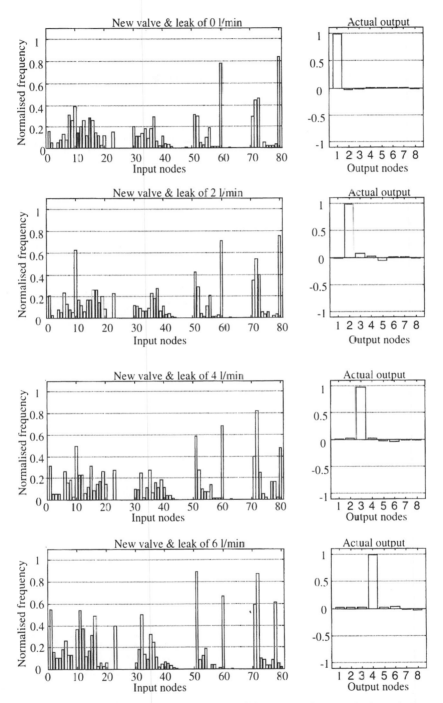

Figure 3.65. ANN performance for the new-condition servovalve [Freebody and Watton, 2000]

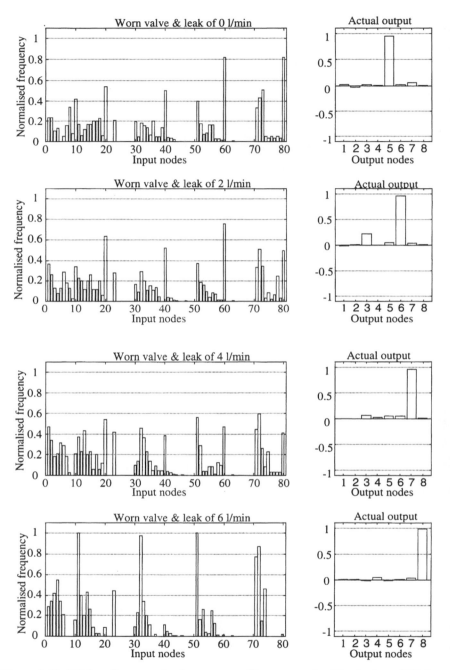

Figure 3.66. ANN performance for the worn-condition servovalve [Freebody and Watton, 2000]

3.5 Further Applications of Artificial Neural Networks

3.5.1 Leakages Within a Position Control System

This system is also considered in Chapter 5 using expert systems concepts on position, and the circuit is shown here as Figure 3.67.

In this application a demand ramp is applied to create a constant actuator velocity and pressure and flow data are collected over the full stroke. Two ANNs are used, one for single faults and the other for multiple faults. A 4–6–3 topology was found to work well and the various faults and associated ANN outputs are as shown in Table 3.4 with the topology for both ANNs illustrated in Figure 3.68.

Table 3.4. ANN output fault logic

Fault	Outputs			
	1	2	3	
No fault	0	0	0	
Line 1	1	0	0	Single faults
Line 2	0	1	0	
Internal	0	0	1	
Line 1 + Line 2	1	1	0	
Line 1 + Internal	1	0	1	Multiple faults
Line 2 + Internal	0	1	1	

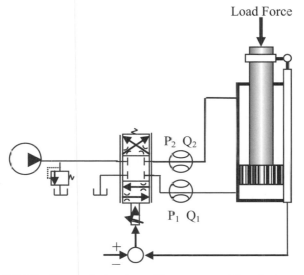

Figure 3.67. Position control system with pressure and flow instrumentation

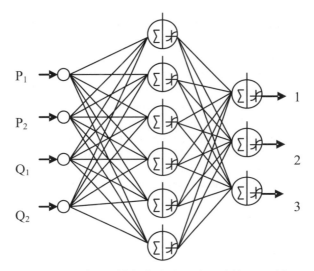

Figure 3.68. ANN topology for multiple fault detection within a position control system

Leakages in the range $0.5 \rightarrow 1$ litre/min were introduced and the following parameters were found to work well:

- Learning coefficient $= 0.1$ Momentum coefficient $= 0.5$
- Sampling frequency 1 kHz, 500 samples per channel used
- 10^6 iterations
- No fault for an output < 0.5 fault for an output > 0.5

Results from several hundred tests, utilising National Instruments DAQ hardware and software, are presented in Table 3.5.

Each 500 point data string for pressures and flow should have constant values for an ideal system. This approach therefore probably works well because practical fluctuations are recorded making training more successful than, for example, from a computer simulation. The procedure, using the trained network, is therefore to

Table 3.5. Results of ANN predictions using new data

	000	001	010	100		011	101	110
000	**226**	8	40	22	011	**254**	13	16
001	2	**226**	–	20	101	43	**341**	–
010	17	–	**206**	18	110	13	–	**302**
100	–	2	4	**209**				
Overall success rate $= 87\%$ Single neuron ANN					Overall success rate $= 90\%$ Multiple neuron ANN			

test for a single fault first and then follow this with a test for multiple faults. Some judgement is needed in some cases where predictions are ambiguous, but this is typical of any ANN application. Further work on valve actuator systems may be studied, for example (Darling and Tilley, 1993). In this approach the supply pressure, two actuator pressures, actuator displacement and time sequence number were used as inputs to the neural network and in the dynamic mode of operation. The neural network had one hidden layer with 60 neurons and an output layer with 10 neurons and training was undertaken for the good condition and for 10 fault conditions. Outputs of 0 for no fault and 1 for a specific fault were then sought. A comparison of just seven of the faults showed good discrimination of each fault at the specified neuron output in a similar manner to the approach outlined.

3.5.2 Weight Changes and the Use of Transmission Line Dynamics to Aid Fault Detection

The pressure control system, previously considered in Section 2.20.5 using transmission line modelling, and Section 3.4.4 using frequency spectra correlation, is again considered. The circuit is shown in Figure 3.69.

The modelling approach in Chapter 2 illustrated a ANN approach based on a mathematical model of the system. Parameters P_1, Q_1, P_2, at the previous sampling interval, were fed into the ANN which then predicted the current value of P_2. If faults are now apparent then the transient data will change and hence the ANN weights will presumably change when the ANN is re-trained. This is the simple

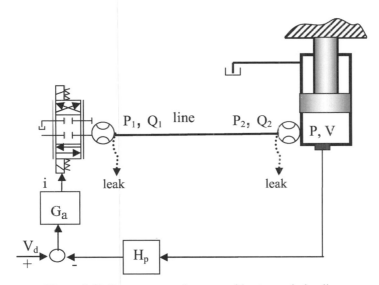

Figure 3.69. Pressure control system with a transmission line

principle of weight change recognition for diagnostic purposes. For this 3-4-1 ANN topology there are 21 weights and biases. Now consider faults as follows:

- A leakage at the actuator end of the line
- A leakage at the servovalve end of the line

Step demanded changes in pressure are applied to the closed-loop system input and the ANN is retrained each time using the transient data and for different leakage flow rates. Some typical transient responses for P_2 are shown in Figure 3.70.

The drop in steady-state pressure is observed for the same leakage flow rate at either end of the line although the transient shapes show no recognisable changes. An important outcome would be the ability to distinguish between leakages at

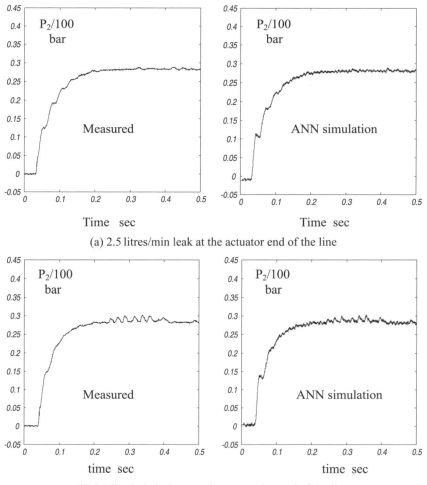

(a) 2.5 litres/min leak at the actuator end of the line

(b) 2.5 litre/min leakage at the servovalve end of the line

Figure 3.70. Pressure transients with and without leakage faults

either end of the line. The ANN re-trained results show good comparisons with the measured data for the two different leakages. In reality it turns out that only the bias b_7 changes by a significant amount, and for leakages at either end of the line or together. This is the bias associated with the pressure P_2 path for neuron 7 and just before the output neuron 8. The results are shown in Figure 3.71.

Further work has shown that a much larger bias increase occurs for smaller actuator volumes, and where transmission line effects therefore become more evident, although again there is no evidence of any other weight changes as initially demonstrated in Figure 3.71. This approach has some value as a back-up technique to support other diagnostic approaches.

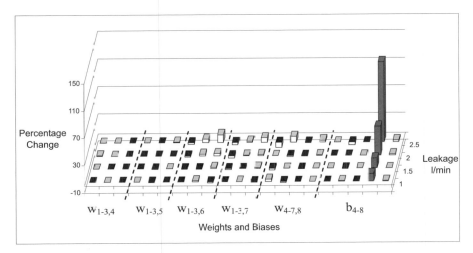

(a) leakages at the actuator end of the line

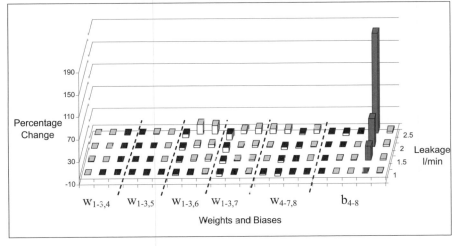

(b) leakages at the servovalve end of the line

Figure 3.71. Weights and bias changes for various leakage conditions

3.5.3 Linear Prediction Coding and Cepstrum (LPC) Features Extraction and ANN Classification

The LPC method is used to represent signal waveforms via a relatively simple calculation method and has been applied quite successfully, for example, for speech analysis. A first-order linear equation is used to represent the present sample x_t and the preceding p samples as follows:

$$x_t + a_1 x_{t-1} + \ldots\ldots\ldots + a_p x_{t-p} = \varepsilon_t \tag{3.33}$$

where ε_t is the prediction error.

This approach is different from the more conventional time series modelling approach since the current output sample prediction is obtained from previous output samples only. No input samples are considered. The approach is therefore output data based rather then model based.

Minimising the total prediction error energy results in a set of p linear equations.

$$\sum_{i=1}^{p} a_i \sum_{t=-\infty}^{\infty} x_{t-i} x_{t-j} = - \sum_{t=-\infty}^{\infty} x_t x_{t-j} \quad 1 \le j \le p \tag{3.34}$$

Assuming $x_t = 0$ for $t < 0$ and $t \ge N$ as a consequence of windowing and defining the autocorrelation of the signal:

$$R(j) = \sum_{t=j}^{t=N-1} x_t x_{t-j} \tag{3.35}$$

$$\sum_{i=1}^{p} a_i R(j-i) = R(j) \quad 1 \le j \le p \tag{3.36}$$

This system of linear equations may be solved using a recursive procedure to yield the predictor coefficients a_i.

This LPC model is also referred to as an all-pole model since it has a system transfer function given by:

$$H(z) = \frac{1}{1 + \sum_{i=1}^{p} a_i z^{-i}} \tag{3.37}$$

Now consider linear systems theory and in particular the output x(t) given as the convolution of the input g(t) and the system impulse response h(t) as follows:

$$x(t) = g(t)\, h(t) \tag{3.38}$$

This is represented by the Fourier Transform equivalents:

$$X(\omega) = G(\omega)\, H(\omega) \tag{3.39}$$

$$\log |X(\omega)| = \log |G(\omega)| + \log |H(\omega)| \tag{3.40}$$

The Cepstrum is defined as the inverse Fourier Transform of $\log |X(\omega)|$ and therefore becomes:

$$c(\tau) = F^{-1} \log |X(\omega)| = F^{-1} \log |G(\omega)| + F^{-1} \log |H(\omega)| \tag{3.41}$$

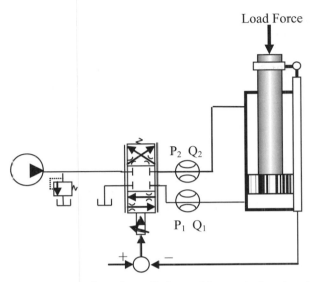

Figure 3.72. Step response testing of a cylinder position control system to obtain LPC cepstra coefficients for fault classification

It can be deduced from (3.41) that the characteristics of the output $x(t)$ and the system $h(t)$ are now separated.

Since τ, often called the quefrency, corresponds to a time domain parameter then $F^{-1} \log |G(\omega)|$ contains high quefrency information and $F^{-1} \log |H(\omega)|$ contains low quefrency information. Therefore the low quefrency components, which are much more easily acquired than the high quefrency components, contain information regarding the system and exactly what is required from a diagnostics point of view.

The Cepstrum may be derived from the LPC coefficients as follows:

$$c_1 = -a_1$$

$$c_n = -a_n - \sum_{m=1}^{n-1}\left(1-\frac{m}{n}\right)a_m c_{n-m} \qquad 1 < n \le p \qquad (3.42)$$

$$c_n = -\sum_{m=1}^{p}\left(1-\frac{m}{n}\right)a_m c_{n-m} \qquad p < n$$

The method will now be applied to the position control system previously discussed in this chapter and which is again shown as Figure 3.72.

Further information is as follows:

- Two step response tests and the two line pressure were used for classification purposes, that is, for both extending and retracting of the cylinder. This is potentially useful in that the approach does not require accurate, and relatively expensive, flow meters.
- Leakages were introduced from each line and across the cylinder seal and having values of 1, 3, 6 litres/min (small, medium, high) in various combinations similar to the previous approach in Section 3.5.1.

- Data was transformed via a Hamming window and a fixed length data capture of 512 points were used although 256, 128, 64 data points were also investigated.
- The sampling rate was 1 kHz and the oil temperature was $45 \pm 5°C$.
- Both extending and retracting information was then used to train an ANN to classify the faults using the first 10 LPC cepstra coefficients. This means that a single test "frame" consists of 40 cepstra coefficients.

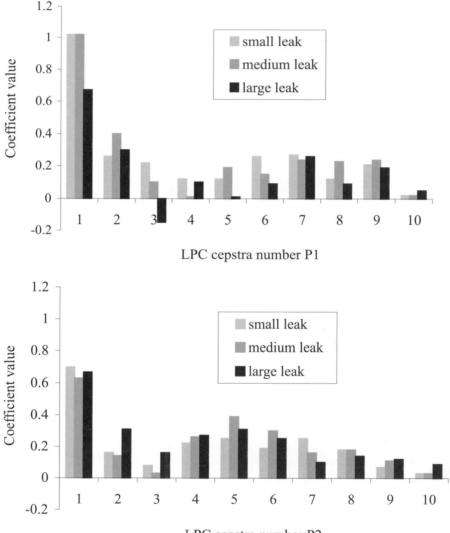

Figure 3.73. First 10 LPC cepstra coefficients for one sampled transient

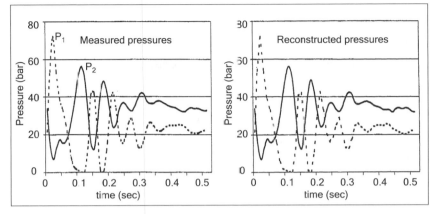

Figure 3.74. Measured and reconstructed pressures using LPC coefficients

Consider just one typical transient response and for the three levels of leakage introduced, in this case from line 2. The first 10 LPC cepstra coefficients for the two line pressures, and their variation with leakage, are shown in Figure 3.73. Also shown in Figure 3.74 are the two typical pressure transients and the reconstructed signals using the LPC coefficients. It can be seen that the method is accurate even for the large variation in pressures experienced during the transient behaviour.

Leakage flows of up to 6 litre/min were introduced and this compares with a maximum transient flow rate of 20 litres/min during testing.

Following a great deal of testing and ANN training it was found the use of the first 10 coefficients gave the highest fault detection accuracy. Training was based on the fault conditions line 1, Line 2, Internal and all combinations of faults. Including the no-leakage condition gives rise to eight no fault/fault/multiple fault conditions, each with three levels of leakage. The next issue is concerned with the way each frame of 40 LPC data points is entered into the ANN. This can be done by entering each frame in series using a number of frames for each fault condition. Alternatively the sampling window can be advanced along the sample axis to provide more frames, each frame then being applied in parallel for a particular transient. Window lengths of 512, 256, 128 and 64 were studied and it was found that a length of 256 gave the best ANN performance. The parallel input approach was marginally better than the series input approach and the optimum topology was found to be 120-50-8. Table 3.6 illustrates the diagnostic accuracy of the approach.

It can be seen from Table 3.6 that the prediction of fault type is always better than 92.7%. A second ANN was also used, operating on similar principles to that previously described, and to predict the actual leakage flow rate magnitude. This was based on the use of flow meters to measure the mean steady-state losses at each fault trial. Again the same topology and data structure as used previously gave the best results and Figure 3.75 illustrates the results obtained. As would be expected, the magnitude prediction becomes very good as the leakage increases. This is a useful aid and certainly becomes useful as a fault progresses. However, the errors are high at the lowest leakage rate of 1 litre/min. Note, however from the previous work that the fault type prediction capability of the technique is excellent.

Table 3.6. Diagnostic type success using LPC Cepstra and an ANN

Leakage type	Identification (%)								Cycles
	None	1	2	i	1,2	1,i	2,i	1,2,i	
None									
Line 1									
Line 2	**100**								50
Line i		**100**							150
Line 1,2		1	**96**		3		1		150
Line 1,i				**99**		1.3			150
Line 1,2,I		3			**93**			4	150
Line 1,2,I						**98**		2	150
							100		150
						1		**99**	150

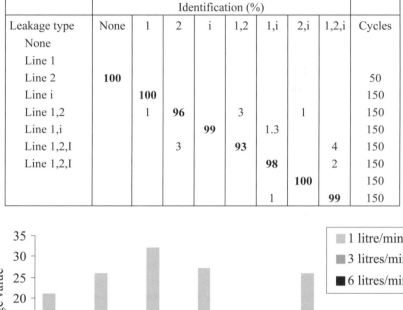

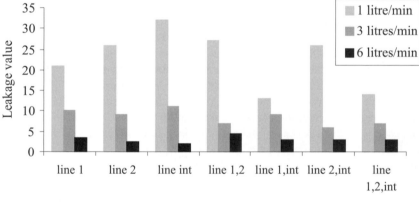

Figure 3.75. Leakage prediction using LPC cepstra coefficients and an ANN

3.5.4 Condition Monitoring of a Bent Axis Pump

Here we consider some of the work carried out by Ramden, (1995) on the application of vibration analysis of a bent axis pump operating at a pressure of 200 bar at 1500 rpm. The following unmodified and fault conditions using faulty components supplied by the pump manufacturer are define as:

(a)   unmodified pump

(b)   two out of three piston rings dismounted from one piston

(c)   a worn outer race for the taper roller bearing

(d)   a worn cut roll in the taper roller bearing

(e)   a worn cut valve plate

An accelerometer was positioned at one of the unused drain ports and accelerometer time signals were sampled at a frequency of 75 kHz with additional signal processing allowing investigations up to a frequency of 10 kHz.

Figure 3.76 shows power spectra for the five conditions illustrating some clearly defined frequency components with magnitudes varying with pump condition.

An artificial neural network was used to train the five conditions and conventional frequency spectra were compared with a new approach using backspectra.

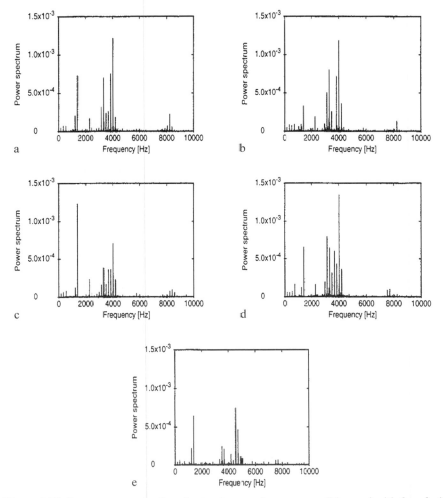

Figure 3.76. Frequency spectra for a bent axis pump in a new condition and with four fault conditions [Ramden, 1995]

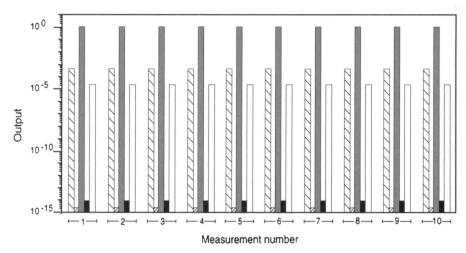

Figure 3.77. Neural network outputs for unseen data with a fault only on the outer race of a taper roller bearing [Ramden, 95]

Both complex and back-propagation methods were used for training although marginally different results were obtained despite advantages and disadvantages of both approaches to spectra and training.

For example, using a 5-4-4-5 topology and using data at 525 Hz, 875 Hz, 1750 Hz, 2975 Hz, 3850 Hz gave excellent fault discrimination. Following appropriate network training for each fault condition, Figure 3.77 shows the identification of fault (c) using unseen data and for 10 measurements.

Note the logarithmic scale on Figure 3.77 and the excellent identification of the fault given the negligibly small values at the other four output neurons.

3.6 Fluid and Wear Debris Analysis

3.6.1 Consideration of the Fluid

A vast range of fluids are available to satisfy different systems requirements and there is no doubt that mineral oil has traditionally been the preferred fluid due to its lubricating properties. However, in many applications such as steel processing and mining, the risk of fire must be minimised and alternative fluids must be used. In addition, environmental considerations are now driving forward the application of water-based fluids for hydraulic applications. Water-based fluids are classified as follows:

HFA 5/95 oil-in-water emulsion, typically 5% oil and 95% water
HFB 60/40 water-in-oil emulsion, typically 60% oil and 40% water
HFC 60/40 water-in-glycol emulsion, typically 60% Glycol and 40% water
HFD Synthetic fluid containing no water

Biodegradable fluids are also receiving attention but it does seem that take-up is slow in this area. However, changing legislation will almost certainly demand that more thought be given to ecologically-friencly fluids. It also seems that in general, components, particularly pumps, can be made to operate with a wide range of fluids available.

The position is improving regarding other components such as servovalves, pressure relief valves, actuators, etc. Water or fluids containing significant proportions of water have been used satisfactorily in certain applications, and despite shortcomings of such hydraulic fluids they have proved to be the only practicable media where massive equipment has to be operated in hazardous environments, and where large volumes of the fluid are required. The apprehension of system users to adopt fluids having a high water content is understandable when the disadvantages of water as a hydraulic medium are considered. Water has negligible lubrication properties, it has relatively high volatility characteristics, it promotes corrosion and can only be used in a relatively restricted range of temperatures. However, water has some distinct advantages when compared to petroleum mineral oil which include fire-resistance, its natural abundance and convenience of high purity supply, its cost effectiveness and its harmlessness with respect to toxicity and environmental pollution.

Type HFA 5/95 oil-in-water emulsions, are fire resistant emulsions which have been designed for fluid power transmission in critical industrial systems. The latest advanced fluids exhibit enhanced stability, lubrication and anti-wear characteristics. The ability of these advanced emulsions to perform as lubricants is due to the combination of their excellent stability towards variations in temperature, pressure, shear and bacterial attack. In addition they provide excellent protection against corrosion and wear, possess low foaming characteristics, can be finely filtered and are broadly compatible with construction materials and seals commonly found in modern hydraulic systems. The performance limitations of conventional 5/95 HFA emulsions became readily apparent for systems operating well above 70 bar, reliability and efficiency often being sacrificed where fire resistance is of paramount importance. Difficulties associated with filterability, emulsion instability and bacterial degradation of these conventional emulsions were not uncommon and contributed to limiting the growth of high-water-based fluids outside of the serious fire hazard situations.

The primary concerns regarding the application of HFA emulsions at high system pressures stem from low fluid viscosity, and the critical effect this has on pump performance, and the relatively poor hydrodynamic lubrication properties of most conventional fluids. In conventional hydraulic pumps a loss of volumetric efficiency is to be expected while operating with a fluid similar in viscosity to that of water (approximately 1 mm2/s @ 40°C). Internal leakage will depend upon the pump configuration: axial and radial piston pumps generally offer the maximum efficiency although certain internal gear pump designs are suitable also. Standard vane pumps and motors are not usually recommended for use with high-water-based fluids, although modified designs are capable of operating satis-factorily at system pressures of around 70 bar.

Category HFB 60/40 water-in-oil emulsions are commonly available although care must be given to the application since the viscosity tends to be much higher than mineral oil particularly at low temperatures. HFC water–glycol fluid contains anti-wear additives and anti-corrosion additives and oxidation inhibitors and makes it suitable for applications in mining, metal processing and other industries where a high fire risk hazard exists. In general, the anti-wear properties of 60/40 water–glycol fluids decrease as the temperature rises, but in some cases the anti-wear properties become more effective with increase in temperature. The anti-wear performance of some fluids can be similar to that of other popular water–glycol fluids, at the higher operating temperatures normally encountered in oil hydraulic systems. Provided appropriate steps are taken to avoid undue loss of water by evaporation, this offers considerable advantages to users.

In general, water–glycol fluids are less effective bearing lubricants than petro-leum mineral hydraulic oils, but are entirely satisfactory in systems containing pumps with plain bearings or lightly loaded ball and roller bearings. However, in common with other water-based fluids a reduction in bearing life can be expected. This will normally be included in the de-rating made by the pump manufacturer.

A synthetic fluid such as a tri-aryl phosphate ester fire-resistant fluid, is blended with additives to give superior characteristics. This type of fluid is desir-able in equipment such as die casting machines, billet loaders, electric arc fur-naces, forging presses and other fire hazard situations. The anti-wear properties of this fluid compare favourably with mineral oils containing anti-wear additives. Compared with a mineral oil of the same kinematic viscosity, phosphate ester fluids have greater dynamic viscosities due to the relative densities, i. e. 1.125 compared with about 0.87 for mineral oils. In some cases phosphate ester fluids are better lubricants than mineral oils. However, where the loading is severe as in rolling bearings and vane pumps, the design should be more conservative.

A comparison of kinematic viscosity for a range of fluids investigated by the author is shown in Figure 3.78.

Figure 3.79 compares the performance of a vane pump operating with this range of fluids, illustrating the effect of fluid viscosity on volumetric efficiency. Note that the vane pump is de-rated in speed and pressure. Also it will be observed that the volumetric efficiency is highest for the 60/40 water-in-oil emulsion.

The operation of directional spool valves on very low viscosity water-based emulsions commonly leads to premature failure by erosion processes. These prob-lems can to some extent be overcome by the selection of more resilient construc-tion materials, such as stainless steel, and designs which avoid conditions of im-pingement erosion by high velocity fluid jets.

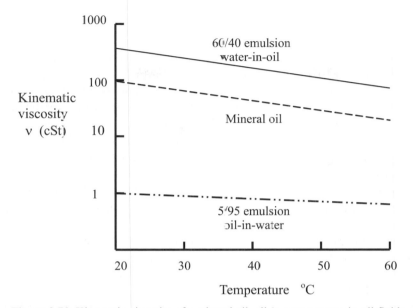

Figure 3.78. Kinematic viscosity of a mineral oil, oil-in-water, water-in-oil fluids

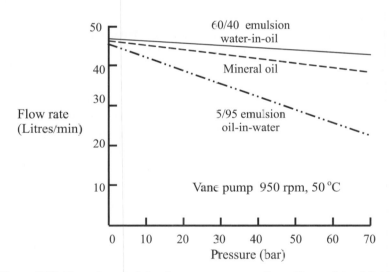

Figure 3.79. Flow characteristic of a vane pump operating with a variety of fluids

The densities and vapour pressure of high-water-based emulsions are notably higher than those typical of petroleum mineral hydraulic oils. It is important to note that the conditions at the pump suction port, which have a strong influence on pump longevity, are affected by fluid density temperature, vapour pressure, pump speed and the static fluid head. The pressure existing at the pump inlet must not fall below the minimum permissible level specified by the pump manufacturer and attention should be paid to the following aspects:

- A positive static fluid head of at least 0.5 m, obtained from a reservoir located above the level of the pump, is usually adequate to provide sufficient suction port pressure, i.e. not less than zero gauge pressure or as required by the manufacturer. Alternatively, a secondary inlet boost pump may be used to obtain the desired conditions. The higher density of water-based fluids compared to those of petroleum mineral hydraulic oils can be problematical where the pump is situated above the fluid level in the reservoir and negative head conditions exist. Failure to maintain the required pump suction port conditions can promote cavitation, erosion and excessive noise. Accelerated wear and malfunction of the system components can result.
- The pipework between the reservoir and the pump suction port should be as short as possible with the minimum number of elbows, bends and restrictions. The bore of the pipework may need to be enlarged if the pump suction port pressure losses cannot be reduced to acceptable levels.
- The recommended pump speed specified by the manufacturer should be strictly adhered to. Increasing pump speed reduces pressure at the suction port and consequently a higher inlet pressure is required if for an acceptable pump life.
- Fine porosity strainers in the suction line should be avoided as they can contribute significantly to inlet pressure losses. Alternative fluid filtration arrangements should be considered, e.g. filter location in the high pressure line and in the return line to the reservoir. The pump suction strainer should be relatively coarse, for example 150 μm–300 μm with a flow rating > 2 times the pump output flow rate being satisfactory. Inert metal mesh strainers are preferred.

Some other key issues to be aware of are as follows:

- High-water-based fluids are less compressible than petroleum mineral oils, i.e. are more "rigid" hydraulic media. As a consequence, faster acting pressure relief valves should be used in order to avoid excessive pressure peaks.
- For reservoirs constructed of mild steel, rusting can occur above the fluid level where condensation can form. Ideally, reservoirs should be constructed of stainless steel although mild steel can be used if suitably protected by a two-pack, chemically-cured epoxy coating. Conventional paint, enamel and varnish finishes invariably show limited compatibility with high-water-based fluids and should be avoided. High-water-based fluids are generally compatible with common construction materials used in hydraulic systems, although magnesium, zinc, cadmium and aluminium metallurgy should be avoided wherever possible.

- High nitrile and neoprene seals and packings recommended for use with petroleum mineral oils are generally satisfactory for use with high-water-based fluids. Fluorinated elastomers, e. g. viton, are also suitable. Polyurethane, butyl, ethylene–propylene, silicone and natural rubber elastomers should be avoided. Packing materials containing asbestos, leather and cork are not recommended since they absorb water. Board and paper material should not be used for flange and cover seals. Fluid packing compounds or mastics should be used sparingly to avoid system contamination and possible blockage of valves and filters.
- Efficient filtration and rigorous contamination control is of great importance when using high-water-based fluids and 10 µm or 15 µm nominal filters should be used, as normally recommended by equipment manufacturers. These should be located in the high pressure line and in the return line to the reservoir after the load valve, or as recommended. The surface of the filters should be large enough to avoid a high pressure drop and the volumetric capacity of all filters should be such that they are able to pass at least three times the output of the pump at the operating viscosity.
- By-passes are not recommended in the high pressure line, and a pressure drop in excess of 3.5 bar is to be avoided. Many types of fluid filters are suitable for use with high-water-based fluids and users should always refer to individual manufacturers' recommendations. Inert metal mesh or glass fibre filters are preferred. Active clay or absorbent filters should not be used. Frequent filter changes are recommended, particularly during the initial stage of operation with high-water-based fluids.

3.6.2 Wear and Particle Contamination

A good circuit design will minimise wear in hydraulic systems, and the addition of attention to fluid cleanliness plays a significant factor in this respect. It is an easy matter to check the particle count for mineral oils but almost impossible for water-based fluids. When one considers the typical clearances and machining tolerances applicable to hydraulic components it should be obvious that some from of contamination monitoring and control should be in place in many applications for reasons outlined in Chapter 1. Table 3.7 shows some typical clearances met in hydraulic control systems.

It will be seen that high pressure filters must be capable of removing particles greater than typically 5 µm. Some common sources of contamination are:

- Built-in contamination *from new components*, for example a reservoir
- Inherent particles in both *new oil* and during *replenishment*
- Ingressed through the *reservoir breather*
- Ingressed through *cylinder rod seals*
- Introduced during maintenance

Table 3.7. Typical clearances for fluid power components [courtesy Star Hydraulics UK]

Component	Typical clearance range (μm)
Bearings	
• Roller	0.1–1
• Journal	0.5–100
• Hydrostatic	1–25
Gears	0.1–5
Dynamic seals	0.05–0.5
Pumps	
Gear	
• Tip to cut in track	0.5–5
• To side plate	0.5–5
• Matching	0.2–5
Vane	
• Tip to cam ring	0.5–1
• Side of vane	5–13
Piston	
• Piston to bore	5–40
• Port plate clearance	0.5–5
Valves	
Directional	2–20
Poppet	10–40
Servo	2–4

There are a number of contamination wear mechanisms, some important ones being:

- *Abrasive wear* due to particles moving between adjacent moving surfaces, particularly within pumps and motors
- *Erosive wear due to particles* moving within a high velocity fluid or jet such as occurs at orifices, spool valve lands, poppet-type seatings
- *Erosive wear due to cavitation*, air bubbles are carried from a low pressure region to a high pressure region causing bubble implosion followed by shock waves that cause the damage
- *Adhesive wear due to metal-to-metal contact*, lubrication films are destroyed when fine clearances are demanded to reduce flow losses
- *Corrosive wear* due to water or chemicals reacting with metallic components, particularly for water-based fluids
- *Water–oil separation* for water-based fluids causing loss of lubricating properties

Cavitation erosion can cause rapid damage, sometimes removing large amounts of metal in minutes for badly designed components. It is often very difficult to avoid high velocity jets due to the small clearances inherent in fluid power components, however, it is resolvable when operating with mineral oil. Some interesting work by (Urata, 2002) has compared cavitation damage on samples of many materials submerged in water, and using an ultrasonic vibrator to create the damage. The main conclusions regarding resistance to erosion were briefly:

- Aluminium bronze was better than stainless steel but phosphor bronze (with a hardness similar to aluminium bronze) was poorer than stainless steel. Aluminium bronze and silicon-A2 were better than titanium alloy. Highly hardened silicon-A2 had a resistance almost 1/3 of the non-hardened alloy.
- Low temperature thermal sprayed ceramics were easily peeled off by cavitation and were considerably weaker than common metals. The coating with diamond-like carbon, DLC, was highly resistive to cavitation damage.
- For resins, hardness was not a significant factor with erosion of soft resins being smaller than that of hard phenol. PTFE showed high erosion properties.
- On the carbon steel specimen, the temperature around the erosion pitting rose up to several hundred degrees centigrade during cavitation attack.

A further detailed study on a vast range of fluids by (Urata, 1998) showed, perhaps as expected, that cavitation erosion is less significant with oil and oil-in-water emulsions than with ionised water. Considering the vane pump test described earlier, and in particular for a 5/95 oil-in-water emulsion, some low temperature tests revealed interesting wear rates which were deduced to be responsible for rapid failures on-plant during winter conditions. The solution of course was to ensure pre-heating of the fluid during such cold conditions to avoid high viscosity effects at low temperatures. Mass loss data for these tests were discussed in Section 3.1.3.

A 20 year study of piston pump life by Pall Corporation, and of a tripper crane at a steel mill showed a four-fold life increase following the fitting of 3 μm clearance protection filters to the hydraulic system. The average pump life evolution is shown in Figure 3.80.

Following this programme some comments may be made:

- Years 1–7 Initial experience of a short pump life
- Years 8–11 System modified to reduce temperature and control ingression (no significant change in filtration)
- Years 12–14 3 μm clearance protection filters added
- Years 15–20 Benefits of clearance protection filtration demonstrated

It has already been said that even new, ostensibly clean, oil may well be contaminated with particles of an unacceptable size. It is therefore important that the contamination specification by the component manufacturer is achieved at the outset to avoid accelerated wear taking place; appropriate filtration should not be an afterthought. Figure 3.81, for example, shows a collection of optical photographs for different fluid conditions.

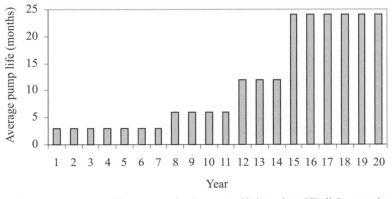

Figure 3.80. Average pump life as contamination control is introduced [Pall Corporation data]

	Description	No particles/millilitre of oil
		>2μm 33,121
		>5μm 7,820
	New oil from barrel	>10μm
	5,010	
		>15μm
	2,440	
		>2μm 79,854
	New system with	>5μm 21,070
	built-in contaminants	>10μm
	12,320	
		>15μm
	8,228	
		>2μm 9,870
	System with typical	>5μm 2,400
	Hydraulic filtration	>10μm 1,800
		>15μm 540
		>2μm 80
	System with $\beta_3 \geq 200$	>5μm 41
	Clearance protection	>10μm 20

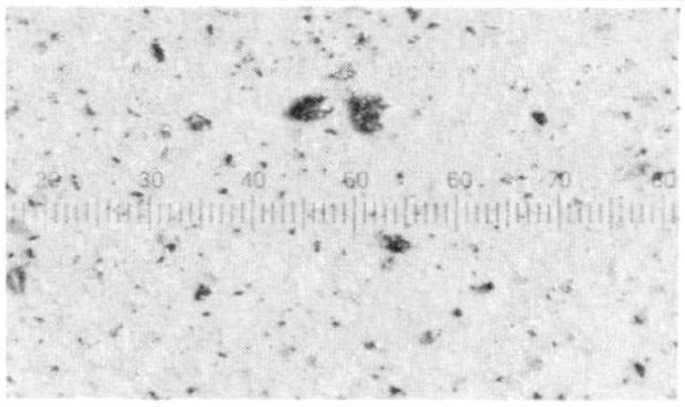

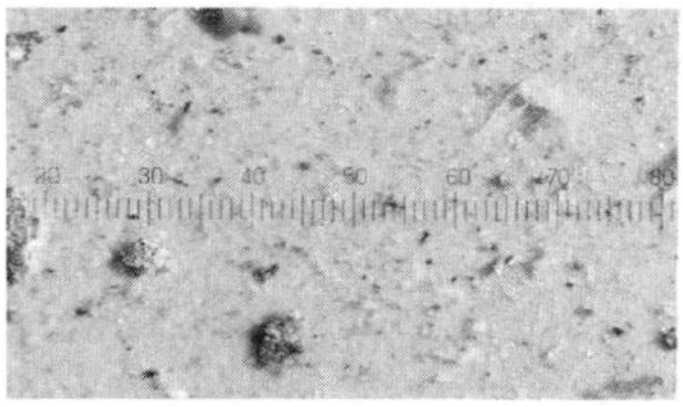

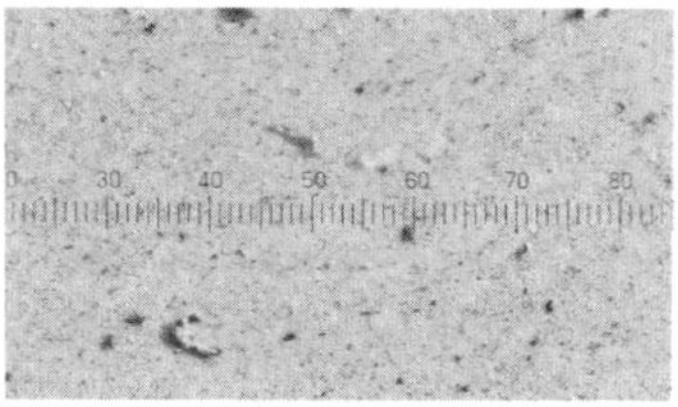

Figure 3.81. Comparison of fluid contaminants for mineral oil in different stages of its working life [Pall Corporation data]

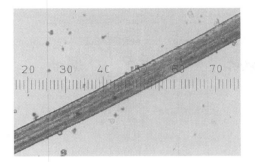

Figure 3.82. Human hair (75 μm), particles (10μm) at 100 x (14 μm/division)

Particle size may be put into context by considering the comparison of around 10μm particles with a human hair as shown in Figure 3.82.

The most widely accepted measure of filter performance is the filtration ratio or beta ratio (β), which is derived from the multi-pass test. This test forms just one of a series of issues that must be addressed for filter selection (cyclic flow, flow surges, pressure pulses, etc.). The β_x ratio is defined quite simply as:

$$\text{Filtration ratio } \beta_x = \frac{\text{Number of upstream particles } x\,\mu\text{m and larger}}{\text{Number of downstream particles } x\,\mu\text{m and larger}} \qquad (3.43)$$

This ratio assumes the same sampling volume upstream and downstream of the filter. For example in Figure 3.81 a system with $\beta_3 \geq 200$ clearance protection filtration is used and is compared with the particle contamination count with conventional filtration, and for a particle size $x = 3$ μm.

3.6.3 Particle Size Classification, ISO4406, Automatic Particle Counters

For more conventional fluid sampling the ISO4406 particle count classification using automatic samplers is now in common use. It provides a reference by which a component cleanliness level may be specified and maintained by the user. Table 3.8 gives the particle size classifications and their range code. To quantify a cleanliness requirement in practice it is then a matter of specifying the required codes at particle sizes of 4 μm, 6 μm, 14 μm. This is usually referred to as 4/6/14(C) indicating a certified count. For applications using microscopes for particle counting or counters pre-1999, two sizes of 5 μm and 15 μm are acceptable and the classification is usually referred to as –/5/15. On pump manufacturer, for example, quotes a cleanliness specification of 17/15/12 indicating a particle count more lenient, as expected, towards the 4 μm and 6 μm sizes as follows:

Code 17   4 μm   640–1300 particles/ml
Code 15   6 μm   160–320 particles/ml
Code 12   14 μm   20–40 particles/ml

Table 3.8. ISO4406 Standard for particle count classification

Number of particles/millilitre oil			Range code
more than		*up to and including*	
80 000	→	160 000	24
40 000	→	80 000	23
20 000	→	40 000	22
10 000	→	20 000	21
5 000	→	10 000	20
2 500	→	5 000	19
1 300	→	2 500	18
640	→	1 300	17
320	→	640	16
160	→	320	15
80	→	160	14
40	→	80	13
20	→	40	12
10	→	20	11
5	→	10	10
2,5	→	5	9
1,3	→	2,5	8
0,64	→	1,3	7
0,32	→	0,64	6
0,16	→	0,32	5

The appropriate size of filter element must then be determined by the absolute minimum size of particle that can be allowed to pass through the pump.

It will be seen from Table 3.8 that each range represents a doubling of the contamination level. The level of control depends upon good system design, good working practice combined with planned monitoring and action where necessary. Considering even the condition of "new" oil, as evident in Figure 3.81, together with potential loss of contaminated oil when changing/repairing components, it might be prudent to have facilities for fluid cleaning such as the equipment shown in Figure 3.83. Such oil filtration systems can also include particle counters, or they may be easily added later if costs justify this. Portable filtration trolleys such as the one shown in Figure 3.83 can handle flow rates up to 45 litres/min and also include simultaneous water removal. The filter trolley shown also incorporates a particle counter, PC 9000, and other trolley models are available from this manufacturer, and able to handle water/sea water removal and for flow rates up to 500 litres/min.

Returning to particle counting, there has been a wide range of instruments developed in recent years, many having now disappeared from the commercial market. The simplest, and often most cost effective, approach is to use a visual technique whereby an oil sample is compared with reference slides via a microscope. This is illustrated in Figure 3.84.

Figure 3.83. Off-line oil filter trolley [www.filtertechnik.co.uk]

This approach to particle counting, sometimes called a patch test, requires sample bottles to be in laboratory clean condition thus requiring some thought to preparation, and often the use of a supply company specializing in such a provision. A good particle count and size assessment can be made using the patch test, and it is also possible to discriminate between different particle materials. This is useful in determining the most probable component that contributes to the wear process. The used bottles must then be carefully re-cleaned to laboratory condition for further use. Many companies also provide an oil analysis service including the supply of sample bottles.

On-line and in-line particle counting represents the most efficient and automated approach when cost justifies its implementation. Figure 3.85, for example, shows such a counter which is also included in the filter trolley shown in Figure 3.83.

Particle counters have been developed principally using either the light blockage method or the filter blockage method, and the counter shown in Figure 3.85 uses the former method. This particular product comes in several forms depending upon the application, for example straight-through, flow diversion, pressure level etc. Data may be transferred via a fibre optic link for remote operation and wireless transmission now forms a basic add-on to instrumentation in general. The light

Figure 3.84. Oil analysis using a microscope/reference slides for particle counting and identification [Pall Corporation]

Figure 3.85. PC9000 particle counter [www.filtertechnik.co.uk]

blockage approach is based on the fact that particles in the fluid block light detected by a photodiode and translated into an electrical signal. The frequency and amplitude of the output signal provides particle size and concentration information. Counters such as the PC9000 produce particle concentration counts in accordance with the appropriate ISO standard and it is proposed that the return on investment can be within the 1–2 year time scale, depending on the frequency of sampling.

3.6.4 Wear Debris Analysis

A more detailed understanding of wear may be obtained by considering the wear particles properties in addition to their size distribution. This requires more advanced equipment such as scanning electron microscopes, image analysis, debris detectors, ferrographic analysis etc [Roylance and Hunt 1999, Evans and Hunt 2003, Radhakrishnan 2003]. These facilities are usually made available through specialist oil analysis companies who undertake planned sampling and analysis followed by a comprehensive report presentation. For example, Table 3.9 shows a report for a powerpack operating with ISO 46 mineral oil and after 600 hours of operation. This powerpack had been routinely monitored and it is clear from the report that action must be taken. In fact, and considering the ISO 4406 cleanliness code, the actual particle count had a trend as shown in Figure 3.86.

Considering another oil contamination study, a microscopic analysis report revealed additional information, and a diagnosis, as shown in Table 3.10.

It is clear from this report that the oil is in a very poor condition and attributed to the oil tank, probably during its construction, together with severe sliding and rolling element wear resulting from the contamination. The material properties are confirmed by the detailed analysis of the particles found, and illustrates the importance of undertaking both particle counting and wear debris analysis to ensure optimum system usage and performance.

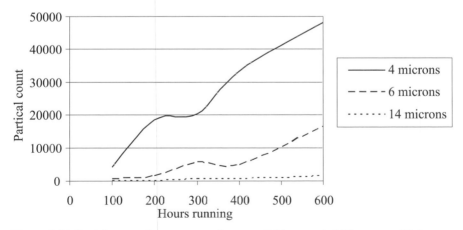

Figure 3.86. Particle count for a powerpack over a 600 hour period [Courtesy of Robertson Laboratories, Conwy UK]

Table 3.9. Oil and wear debris analysis for a powerpack [Courtesy of Robertson Laboratories, Conwy UK]

Robertson Laboratories
Conwy
LL32 8FA
United Kingdom
Tel: 01492 574750
Fax: 01492 574778

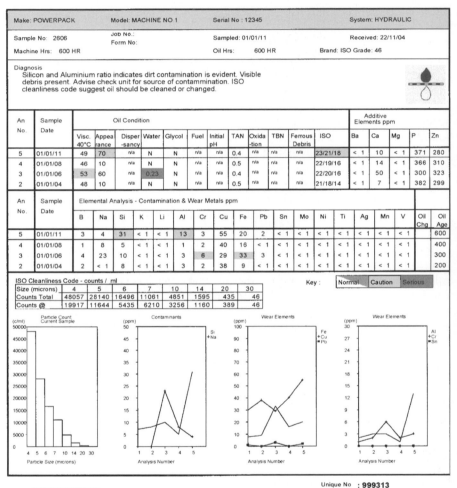

Make: POWERPACK	Model: MACHINE NO 1	Serial No : 12345	System: HYDRAULIC
Sample No: 2606	Job No.: / Form No:	Sampled: 01/01/11	Received: 22/11/04
Machine Hrs: 600 HR		Oil Hrs: 600 HR	Brand: ISO Grade: 46

Diagnosis
Silicon and Aluminium ratio indicates dirt contamination is evident. Visible debris present. Advise check unit for source of contammination. ISO cleanliness code suggest oil should be cleaned or changed.

An No.	Sample Date	Visc. 40°C	Appearance	Disper-sancy	Water	Glycol	Fuel	Initial pH	TAN	Oxida-tion	TBN	Ferrous Debris	ISO	Ba	Ca	Mg	P	Zn
5	01/01/11	49	70	n/a	N	N	n/a	n/a	0.4	n/a	n/a	n/a	23/21/18	< 1	10	< 1	371	280
4	01/01/08	46	10	n/a	N	N	n/a	n/a	0.5	n/a	n/a	n/a	22/19/16	< 1	14	< 1	366	310
3	01/01/06	53	60	n/a	0.23	N	n/a	n/a	0.4	n/a	n/a	n/a	22/20/16	< 1	50	< 1	300	323
2	01/01/04	48	10	n/a	N	N	n/a	n/a	0.5	n/a	n/a	n/a	21/18/14	< 1	7	< 1	382	299

Elemental Analysis - Contamination & Wear Metals ppm

An No.	Sample Date	B	Na	Si	K	Li	Al	Cr	Cu	Fe	Pb	Sn	Mo	Ni	Ti	Ag	Mn	V	Oil Chg	Oil Age
5	01/01/11	3	4	31	< 1	< 1	13	3	55	20	2	< 1	< 1	< 1	< 1	< 1	< 1	< 1		600
4	01/01/08	1	8	5	< 1	< 1	1	2	40	16	< 1	< 1	< 1	< 1	< 1	< 1	< 1	< 1		400
3	01/01/06	4	23	10	< 1	< 1	3	6	29	33	3	< 1	< 1	< 1	< 1	< 1	< 1	< 1		300
2	01/01/04	2	< 1	8	< 1	< 1	3	2	38	9	< 1	< 1	< 1	< 1	< 1	< 1	< 1	< 1		200

ISO Cleanliness Code - counts / ml

Size (microns)	4	5	6	7	10	14	20	30
Counts Total	48057	28140	16496	11061	4851	1595	435	46
Counts @	19917	11644	5435	6210	3256	1160	389	46

Key : Normal | Caution | Serious

CARDIFF UNIVERSITY
ENGINEERING BUILDING
CARDIFF
SOUTH GLAMORGAN
CF1 1AB

Unique No : **999313**
Signed   : Gwyn Simmonds
Reference : 688222/REF SAMPLES/1/RL/H
Date   : 22/11/2004 16:42:16
Location: *

Table 3.10. Microscopic analysis report [Courtesy of Robertson Laboratories, Conwy UK]

MICROSCOPIC ANALYSIS REPORT

Sample No.: **1630428**	Serial No. : **P100/5827**	Unit Age: **1 HOUR**	System: **HYDRAULIC**
Order No. : **5221**	Model: NOT GIVEN	Location: **BOTTOM OF TANK**	Oil Brand: NOT GIVEN

REASON FOR ANALYSIS: **Microscopic Testing**

	DEBRIS IDENTIFICATION				PHYSICAL TEST DATA	
% TOTAL	DIRT/ GRIT	METAL	SHOT BLAST/ WELDING SLAG	FIBRES/ PLASTIC		
	65	20	5	10		
SIZE (uM)						
>5	>99	>99	>99	>99	Kv @40'C	48
>10	60	65	50	75	APPEARANCE	VISIBLE DEBRIS
>25	15	25	20	35	WATER %	<0.1
>50	5	15	<1	10	ISO CODE	23/21/16

ELEMENTAL ANALYSIS (ppm) DETERMINED ON OIL														
NA	SI	AL	CR	CU	FE	PB	SN	MO	CA	MG	P	B	ZN	K
<1	3	<1	<1	<1	6	<1	2	2	38	1	322	<1	401	<1

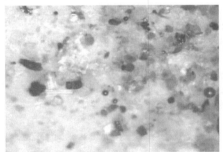

MICROSCOPIC IMAGE X200   MICROSCOPIC IMAGE X100

DIAGNOSIS

No oil data stated. Routine laboratory analysis indicated a high ISO code and visible debris.
Microscopic analysis of the debris indicated a high level of dirt and metal. Much of the metal debris was consistent with abnormal wear as shown in the right hand image. The type of wear detected was typical of severe sliding wear and rolling element wear. Many of the wear particles also showed evidence of blue temper colours indicative of very high temperatures. Shot Blast/Welding slag was also present and most likely to be derived from manufacture of the oil tank. The contamination detected, namely Dirt and Shot Blast Welding Slag would contribute significantly to the abnormal wear processes evident. Over 99% of the debris would pass through a 125 micron strainer. Although this sample may not be representative as stated in your covering letter it is recommended that the unit should be checked for abnormal noise or heat due to the severity of wear detected. A follow up representative sample from the system to monitor overall wear and cleanliness levels is also recommended.

F.A.O. Mr J Watton
Cardiff Univeristy

Signed   : Gwynne Thomas
Reference : 1630428
Date   : 08/04/02

KEY	NORMAL	CAUTION	SERIOUS

3.7 Temperature Sensing

Temperature measurement is perhaps the simplest and lowest cost form of monitoring. It is desirable to keep the operating temperature of a fluid power system within a specified working range and closed-loop temperature control may be achieved with a degree of success using either a water cooler or fan-driven cooler in the system main return line.

From a condition monitoring point of view changes in temperature, for example in a pump case drain line, may indicate a change in leakage flow rate and hence wear of a component within the pump. The change in temperature differential across a component may also indicate component deterioration, but this and most other applications require the measurement of small temperature differential changes. Temperature probes may easily be fitted into lines and components, and the most common type utilise the thermocouple principle of combining dissimilar metals which generates an emf change with temperature change at the junction. Platinum resistance thermometers (PRTs) utilise resistance change to generate a voltage signal and are commonly used in fluid power systems applications, although semiconductor devices utilising thin film sensing resistors are now available at a relatively low cost.

The steady-state temperature drop across fluid power components and its variation with component deterioration poses a difficult analytical problem due to the complex heat transfer processes that exist, and further work is needed in this area. Increasing fluid mass flow rate across a component will normally give rise to increasing temperature differential changes but the heat transfer from the fluid through the component body will also change. Hence the complex interaction between convective and conductive heat flow can produce highly nonlinear temperature changes.

Consider for example Figure 3.87 which shows the temperature differential measurement across a relief valve, that is the outlet temperature minus the inlet temperature $(T_o–T_i)$. PRT temperature probes were used and measurements were taken for different cracking pressures and flow rates through the valve.

Above flow rates of 15 litres/min the temperature differential increases with increasing flow rate and the rate of change of temperature differential is reasonably constant. The effect of increasing the relief valve setting is more distinct and it would appear as though the technique might be of value in detecting such changes providing the flow rate is known. The use of such a measurement for detecting faults within the valve is perhaps limited since the operating conditions must be known and small temperature changes are being compared. Changes in temperature usually occur with a first-order dynamic characteristic and the large system time constant must also take into account.

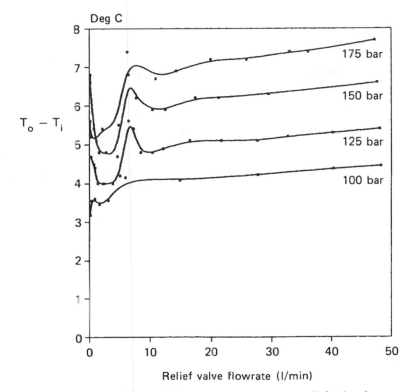

Figure 3.87. Temperature differential change across a pressure relief valve for a range of settings and flow rates through the valve

3.8 Data Acquisition

3.8.1 Hand-held Equipment

As discussed briefly in Chapter 1, there are a number of approaches to data acquisition, the solution adopted depending upon the application complexity and the payback period desired. The most common plant measurements often involve vibration/noise analysis via accelerometers or pressure transducers. This usually calls upon Fast Fourier Transform frequency spectrum analysis although mean level changes can often be useful. If transient signal analysis is preferred then, as with FFT analysis, fast data sampling is still essential. This is not a particularly difficult or expensive issue since there are a variety of instruments available with adequate computing and data storage features. For example, Figure 3.88 shows one type of hand-held data collector with powerful signal processing and diagnostic capabilities for two channels of data.

Figure 3.88. Two-channel hand-held data acquisition and condition monitoring instrument DI440 [www.diagnostic-instruments.com]

This instrument is just one of several available to cover a wide range of applications with appropriate diagnostic routines and graphics display. Some further key features worth noting are:

- It has two signal inputs with a frequency range from DC to 40 kHz
- Comparisons with vibration standards library (ISO, BS, ANSI)
- Comprehensive FFT analysis (100–12,800 lines), bump test analysis for structures natural frequencies, crack detection, structural machine integrity
- User defined modules for custom conformance standards
- Off-line analysis and reporting
- 1 or 2 plane balancing, run-up/steady-state/run-down analysis
- Bearing analysis using ESP^{TM}
- Display real images of machinery and transducer positions
- Comprehensive data logging and communication protocols, digital signal recorder allowing live dynamic data to be recorded and stored for playback at a later stage
- Any sensor signals possible and stored as a standard wave file
- It is a rugged device with a 2 metre multi-drop rating
- Data integrity assurance even in the event of a complete system reset
- An analysis package can be tailored to suit specific applications

3.8.2 Distributed Sensing, Wireless Communication

For a more permanent installation on plant requiring many measurements on a routine basis, an approach such as shown in Figure 3.89 represents one of the latest developments in wireless communication.

Figure 3.89. WiSNet wireless sensor network for monitoring [Expert Monitoring UK, www.expertmon.com]

An overview of the WiSNet® Wireless Sensor Network

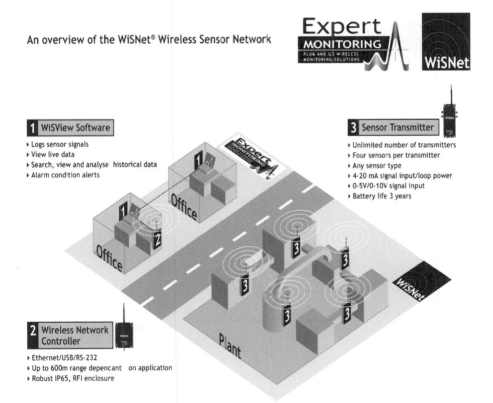

1 WiSView Software

▸ Logs sensor signals
▸ View live data
▸ Search, view and analyse historical data
▸ Alarm condition alerts

3 Sensor Transmitter

▸ Unlimited number of transmitters
▸ Four sensors per transmitter
▸ Any sensor type
▸ 4-20 mA signal input/loop power
▸ 0-5V/0-10V signal input
▸ Battery life 3 years

2 Wireless Network Controller

▸ Ethernet/USB/RS-232
▸ Up to 600m range depencant on application
▸ Robust IP65, RFI enclosure

Figure 3.90. Plant-wide monitoring using wireless network controller [Expert Monitoring UK, www.expertmon.com]

A typical WiSNet system consists of an IP65 Ethernet network controller and sensor transmitter modules that facilitates ad-hoc wireless sensor networks to enable sensors to be positioned anywhere inside or outside of any manufacturing plant. The individual WiSNet wireless transmitter modules are positioned plant-wide, where each module powers any four sensors, and offers a substantial reduction in installation costs, often a matter of minutes rather than days. This is indicated in Figure 3.90.

The WiSNet product utilises a long range (1.2 km LOS) Bluetooth digital radio interface which is plug and play that makes this wireless product immune to both electromagnetic and radio interference. The wireless receiver hub serves up to 255 transmitter modules and the integral viewer software configures, logs and manages all wireless network data and events with minimal user interaction. It is then a simple matter to search, record, plot and analyse historical data and events and also create alarm alerts and event analysis. This system is a cost effective approach to modern system condition monitoring where trend data needs to be established.

Anyone working on a large plant with hard-wired sensors will appreciate the advantage of a wireless sensor system and its elimination of repair costs due to cable breakages that inevitably occur during operation. The system also has a further advantage of remote monitoring in that there is flexibility in the placement and/or removal of the monitoring computer.

3.8.3 Data Acquisition Cards for System Development

If the intention is to develop a monitoring system using a data acquisition card (DAQ) and commercial software to be customized to a specific application, then the DAQ shown in Figure 3.91 is one of many hardware possibilities and system configurations available. This particular card is a high-accuracy multifunction card offering 18 bit resolution, up to 32 analogue inputs, up to four analogue outputs at 16 bit resolution, programmable signal ranges, up to 48 digital I/O lines, two 32 bit 80 MHz counter/timers, analogue and digital triggering, 6 DMA channels for high speed data throughput, sampling rates up to 625 kHz. These high-performance DAQ cards are more than capable of handling typical hydraulic systems data. Lower-cost cards from the same manufacturer and with 12 bit resolution, as used for the practical examples in this book, are quite acceptable.

The DAQ card is placed in one of the expansion slots in the monitoring computer and the accompanying software rapidly configures the card for use. The National Instruments LabVIEW DAQ software may then be incorporated to provide an advanced development system that may then be used on the plant to be monitored. For a large number of channels, and industry-compatible equipment, the PXI/PCI modular instrumentation approach is far more flexible.

The National Instruments DAQ software, LabVIEW, now offers a structured block diagram graphics approach to setting up the data acquisition, calling the

Figure 3.91. M series data acquisition card [National Instruments ni.com/uk]

appropriate signal channel and performing signal processing and graphics presentation of the information that represents the plant condition. An example is shown in Figure 3.92 which illustrates an FFT measurement, and a typical result for a pump pressure and vibration measurement was shown in Section 3.3.

Consider an application on a steel rolling mill which is ideally suited to the development of a bespoke monitoring system based upon National Instruments hardware and software.

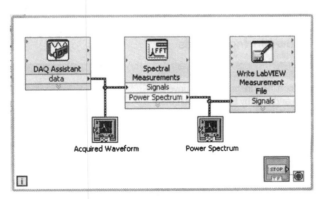

Figure 3.92. Setting up a measurement using National Instruments LabVIEW 7.1 DAQ software

3.8.4 Application to a Hot Steel Strip Finishing Mill

A hot strip finishing mill is illustrated schematically in Figure 3.93. Billets of steel, produced from iron ore, arrive at the mill and are kept in a reheat furnace to ensure that the correct material properties are maintained throughout the remaining processes. The steel slabs then pass through a reversing roughing mill where the majority of scale reduction is performed. The steel strips are then coiled, de-coiled, and then passed onto the finishing mill stands (actually labelled as stands F5–F11). The strip is then water-cooled, coiled, and ready for further processing in the cold mill. The steel strip is gradually reduced in thickness as it passes through each stand.

In order to achieve this and other shape and profile characteristics there are three major hydraulic control systems present. They are specified as:

- Work Roll Bending   (WRB)
- Work Roll Shift   (WRS)
- Automatic Gauge Control   (AGC)

Pressure, servovalve current, and position transducers within these systems continuously send recorded information back to the control station, in order to reduce shape and profile errors and maintain strip quality. Before a condition monitoring system of any sort could be designed, installed and implemented on the finishing mill, a solid base for development had to be built.

The main objective was to successfully introduce a robust on-line fault detection system for the first of the three major hydraulic control systems, the WRB system. The Networked Computer Architecture is shown as Figure 3.94 and the GUI developed within National Instruments LabVIEW is shown as Figure 3.95.

The high-speed feature extraction software controls data acquisition while also sorting and classifying faults into data files. Within the on-line system, faults are generally specified as being mechanical, electrical or hydraulic, and are available within the networked system for clear recognition by the user. All pressure, flow and servovalve transducer signals are opto-isolated to ensure plant operating integrity and an FFT analysis is continually performed on these signals.

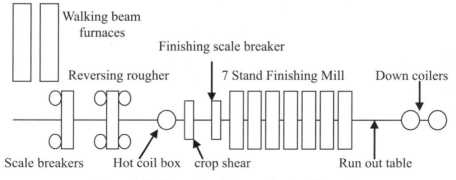

Figure 3.93. Schematic of a hot steel strip rolling mill

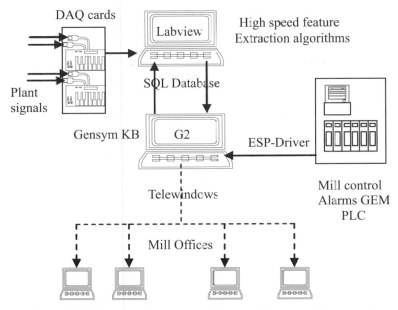

Figure 3.94. Networked computer architecture [Courtesy of Corus plc]

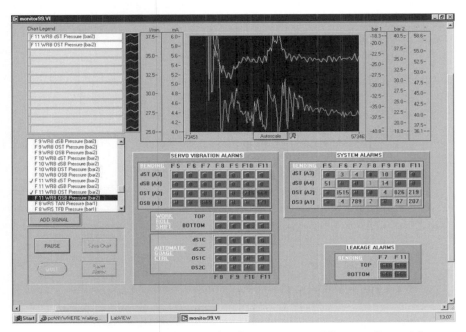

Figure 3.95. Data acquisition GUI for plant operator use [Courtesy Corus plc]

The hydraulic signals for 28 servovalves are continually monitored and a visually-clear alarm notation is used to indicate the condition of each servovalve on each stand via analysis of the frequency spectra taken during the rolling process triggered by the entry of the steel strip at the appropriate stand. The sampling frequency was set at 1 kHz enabling the maximum frequency of interest, 450 Hz, to be identified.

This approach has significantly reduced mill downtime by a more structured consideration of servovalve refurbishment/replacement together with added information on the general rolling performance characteristic. The best estimate suggests that the investment has paid back within two years, given the high cost of mill downtime on this and any other rolling mill system.

3.9 Expert Systems and Knowledge-based Reasoning

3.9.1 What is an Expert System?

- An expert system is a computer program that provides an interface between an expert in any particular field and the processing power of a modern computer.
- It is capable of replicating the problem solving behaviour of human expertise within a known domain and can be used in any area where expertise and experience can be broken down into a set of logical rules.
- The more complex the task the more rules needed to solve it, the more can be gained by using an expert system. This knowledge is utilised to provide decision support at a level comparable to the human expert and is capable of justifying its reasoning by use of a variety of checks and balances.
- Expert systems may be applied to a number of problems including those without engineering science awareness but in the context of engineering, expert systems have been applied to a variety of areas such as:
    - (i)   Planning and management
    - (ii)   Systems design, static, dynamic, component selection
    - (iii) Condition monitoring and fault diagnosis
    - (iv)   Real-time control
    - (v)   Education and training
    - (vi)   Environmental and pollution control
    - (vii) Debugging

The process of building an expert system has often been referred to as knowledge engineering, whereby substantial problems may be solved when explicit knowledge may be readily formulated. So, if an expert system is no more than a computer program, why is it different from a traditional computer program? The answer to this may be explained as follows:

(i) A traditional computer program does not have flexibility with respect to the addition of new knowledge and consequently the way it integrates the knowledge. In an expert system the relevant expertise is held in a knowledge base. This consists of a number of rules that are written in a language that allows them to be read and understood almost like everyday speech. Furthermore the rules can be placed in absolutely any order and there are no loops or subroutines to be concerned with; each rule stands on its own. These facts make it easy for a person with no programming experience to write a new knowledge base or to understand and add to an existing one. With other higher-level computer languages it is often impossible, even for an experienced programmer, to understand another's code.

(ii) The rules may be symbolic or numerical in marked contrast to traditional evaluation of equations or data. The knowledge base exists as a separate entity and requires programs, that can retrieve and manipulate the knowledge it contains. An inference engine interprets the knowledge base and results are recorded in a database. This process, called consultation, is the solving of the required task and once begun is controlled by the expert system. During a consultation the expert system may need information from the user, in which case it will ask questions structured so that the user must apply the missing data. These questions can be written into the application or else will be automatically generated during the inferencing process. Data can also be included as part of the knowledge base or retrieved from other programs.

(iii) An expert system is able to handle information which may not be precisely known, non-deterministic, and with a reasoning uncertainty. In some instances such as on-line condition monitoring the expert system may be able to deduce transducer integrity, that is, is the transducer measuring correctly?

So, in principle, an expert system consists of:

- a knowledge base (KB), which contains the problem solving know-how
- a rule interpreter (inference engine) able to process the knowledge base
- a database, which holds the current status of consulting a knowledge base

This may be illustrated simplistically, as shown in Figure 3.96.

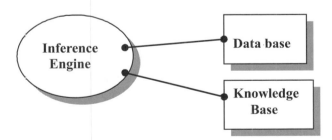

Figure 3.96. Expert system fundamental components

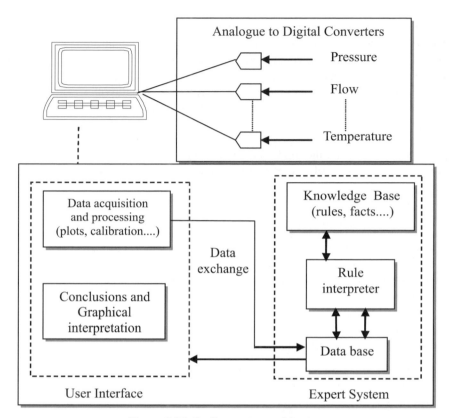

Figure 3.97. On-line system architecture

In reality engineers often need to embrace other aspects such as on-line data acquisition, processing and outputting, and visual graphical communication. A Windows environment allows design of a graphical user interface (GUI) that can integrate all the features mentioned, as shown in Figure 3.97.

An expert system (set of disks and manuals) is often referred to as an expert system shell, since it is, of course, not problem specific and will be empty of rules! The latest expert system shells run in Windows and hence can make use of pull-down menus, dynamic data exchange (DDE), and Bitmaps / pictures / video clips, etc. to represent information in a graphical form of value to the system user in an engineering application.

3.9.2 Generating a Knowledge Base

When initiating a project, a new application is opened, named and assigned to a file. At the beginning of subsequent sessions the application name can be selected from a menu. An application can consist of one or more knowledge bases that contain the actual know-how of any expert system. A new knowledge base is started by selecting text from the edit menu whereby a blank page is displayed onto which rules can be written. Once edited, a knowledge base can be saved to a file and named so that it can be loaded in the future for editing or consultation.

Although there are a variety of expert systems commercially available, the knowledge base is usually made up of rules, demons, facts, defaults and comments written to resemble spoken phrases.

(i) Rules are the building blocks of a knowledge base and can be placed in any order. A rule consists of one or more conditions and one or more consequences, for example:

| If | {temperature is high} | .... | condition |
| then | {action is stop} | .... | consequence |

If the conditions are true then the rule is fired , i. e. the consequences are made true. The keyword to begin a condition is "if", and to begin a consequence is then. Further conditions and consequences have the keyword, and alternative conditions have the keyword or.

if	{temperature is high} or {pressure is high}	....	condition
or if	{vibration is high}	....	condition
and	{strip is out of gauge}	....	condition
then	{action is stop}	....	consequence
and	{check hydraulic circuit}	....	consequence

The fifth rule is an example of an assertion. Whereas most lines consist of an identifier {temperature}, a relation { is }, and a value { high }, an assertion is a complete phrase treated as either true or false.

(ii) A demon is a priority rule that has the keyword "when" and is fired immediately its condition becomes true. It is particularly useful for warnings and/or critical safety conditions etc:

| when | {temperature > 70} |
| then | {picture name is warning} |

(iii) A fact is a rule to unconditionally assign a value to an identifier:

| fact | {colour is green} |
| fact | {force is $P1 \times A1 - P2 \times A2$} |

(iv) A default rule is only fired when the expert system cannot establish a value for an identifier from any other rule:

default   {action is continue}

(v) A comment is an expression inserted to help the understanding of the knowledge base, but which plays no part in the inferencing process.

3.9.3 The Inferencing Process

The way in which the inferencing of a KB is controlled is an important factor to the designer of an expert system application. A user runs the expert system by starting a consultation in one of two ways and this determines what type of inferencing is invoked. Expert systems support two types of inferencing: backward chaining and forward chaining, the former being the most commonly used.

- If a KB item is queried, backward chaining begins to evaluate this goal.
- If data is volunteered then this new data is used for forward chaining.

Backward Chaining

Backward chaining begins from a goal, which is most likely a query initiated by the user. The purpose of backward chaining is to evaluate this goal and to do this it will find a rule containing the goal in its conclusions and try to satisfy the conditions of the rule. If the conditions are satisfied then the conclusion is true; a value for the query item has been found and the consultation is complete.

If the conditions are not known, other rules may have to be examined, creating a chain of linked rules. The chain ends at a rule whose condition is known or can be determined from the user.

In the following example, the required goal might be a value for the identifier action and to start a consultation the user would start a query for action. This can be done from the consult menu or by typing "action?" from the main screen.

if	temperature is high
or if	vibration is high
then	action is stop
if	strip is in gauge
and	temperature is normal
and	vibration is low
then	action is go
if	strip is out of gauge
and	temperature is low
and	vibration is low
then	action is stop
default	action is continue

The backward chaining process would then find a rule with a value for action in its conclusions, such as the first rule (note that it might equally well go to the second rule first). To fire the first rule and thereby make the conclusion true – action is stop – one or other of the conditions must be true, i.e. temperature is high or vibration is high. If a value for temperature, for example, is not already known to the expert system, i.e. it is not in the database, and cannot be found from other rules, then a suitable question will be automatically generated and displayed to the user with possible answers:

temperature is?

Select one of...

high
medium
low

The user is able to select an answer with cursor keys, which is then recorded in the database. If temperature or vibration is selected, the rule will fire and a value for the query item, action, will have been found, thus ending the consultation. If not, then another rule with action in the conclusions will be found and its conditions checked. If no value for action can be found by any other means then the default rule is fired and the result of the consultation will be action to continue.

Forward Chaining

Forward chaining begins with a new item of data, which can be volunteered by the user or may be the consequence of a rule that has just fired. The inferencing process works through the rules and demons to find the implications of this new data. In the same example, a forward chaining consultation could be started by volunteering some data, such as "temperature is high". This can be done in the consult menu or by simply typing the statement from the main screen. The inferencing process then forward chains through each rule until the volunteered data causes a rule to fire, in this case the first rule, so action is stop. This new data is added to the database and may in turn cause other rules to fire, and so on. Normally, backward chaining is used to evaluate a goal and forward chaining is involved only in the use of demons ('when' rules) and not in the more common 'if' rules. When a rule fires, the consequences are forward chained to existing demons which will fire if the new data satisfies their conditions. The two forms of inferencing interact smoothly and without consultation propagator intervention. A knowledge base can also be set to allow forward chaining on rules as well as demons, so that if data are volunteered, only forward chaining will occur.

3.9.4 Supporting Software

It has been shown that expert systems are well structured to deal with typical applications where expertise can be expressed in a form allowing the problem to be broken down into a series of questions that can be answered by a less experienced user. This simple method of placing expertise into a knowledge base has more recently encouraged less typical uses inclucing on-line applications. However, in order to adapt to such functions, expert systems have to rely on support from more flexible and established software.

An obvious requirement for any on-line computing system is data acquisition. To get real values from the real world requires an analogue to digital converter (A/D) to read voltages and transducers. Nowadays this is typically an A/D card that plugs into a PC expansion slot. Software is then required to control the A/D card's functions, such as configuration, number of channels, triggering, sample size, sample rate, gain, etc. Some A/D cards are supplied with simple data acquisition programs (which are best used just to test the cards) and there are commercial packages readily available. When considering flexibility (and cost) it is usually necessary to write customised acquisition routines. To this effect, A/D cards come with functions and examples for a number of programming languages such as BASIC, FORTRAN or C.

As well as retrieving the raw data from transducers, it is always necessary to do some kind of processing of the data before it is useful. Once acquired, data is stored in one long buffer of either integers or voltages which are totally meaningless until calibrated to flow rates, temperatures and forces, etc., followed by further processing. Expert systems are capable of incorporating arithmetic, but without conventional arrays or loops it would be totally unfeasible to present an expert system with 30,000 values – calibrated or not. Instead the expert system must be given, for example peak flowrate, average temperature or total work done, requiring further processing from a high level language.

If any kind of graphical representation is desirable in a system, it is better dealt with by subsidiary software. The expert system may have provisions for displaying pictures and building customised forms for gathering data and displaying outputs, which may be enough for many applications. But if graphing is required or a combination of illustration and text, further software is needed.

3.9.5 Knowledge Elicitation

Often mathematically rigorous descriptions are difficult to solve and must then be characterised by the expert who experiences a "gut feel" about the problem. Knowledge elicitation can present difficulties when approaching a new area of interest, and some effort has been made in characterising the process. For example, the choice of "expert" gives rise to a number of aspects which must be taken into account, such as:

- the sceptical expert
- the inarticulate expert
- the expert who fears replacement or redundancy
- the non-committed expert
- the out-of-date expert/administrator
- the inaccessible expert

- the expert without expertise
- the salesperson!

The choice of more than one expert has advantages such as:

- increased accessibility
- alternative source for explanations
- different approaches to the problem are available
- less chance of missing vital information
- elicitation of exceptions to the rule
- bias counterbalanced
- saving in project time
- rapid verification

However, some disadvantages are:

- contrasting information
- questioning of integrity
- covering same ground several times

The choice of an expert/experts may therefore not be a simple matter and leads to thought being given to the following:

- choice of methodology
- assessing the experts
- interviewing techniques
- the importance of feedback
- analysis and description of elicited knowledge

The examples given previously relate to empirically derived information sometimes called heuristic knowledge, and expert systems containing such rules are referred to as shallow reasoning systems. They do have limitations:

- they cannot accommodate new or unexpected situations
- from the previous comments on knowledge elicitation it is possible that not all the heuristic knowledge has been gathered.

An alternative approach is therefore to adopt a deep reasoning strategy which embraces a mathematical model of the system. This approach more than likely requires a detailed knowledge of the system behaviour including both steady-state and dynamic performance characteristics. These models could be qualitative or quantitative or a mixture of both. In practice a compromise may have to be made due to the complexity of industrial systems, and a hybrid approach combining both shallow reasoning and deep reasoning may prove effective as will be demonstrated in the following section of this chapter.

3.9.6 Working Examples

Fault diagnosis in fluid systems will inevitably include a deep reasoning approach and might combine one or more of several aids such as:

- the known system operating characteristics which have been measured and represented by a sufficiently accurate set of equations
- the use of fundamental mathematical models, fault deductions from comprehensive CAD studies, etc.

Whatever the base-line model used, the approach is to compare measured data with expected data and to deduce the type of fault that has caused the change. It is very rare that precise numbers can be given to a reference model and some acceptable tolerance must be given to the comparisons made. This tolerance arises from uncertainties in the reference data, measured data, and instrumentation integrity.

Example 1. Circuit Flow Monitoring – Some Basic Concepts

Consider the proposition to use three flow meters to ostensibly check the condition of each of the circuit components, that is, the pump, pressure relief valve and servovalve as shown in Figure 3.98.

Assuming all the system parameters are measured (pump speed, supply pressure, all flow rates, servovalve load pressure, servovalve current) then each component characteristic can be compared with the expected characteristic from the database of equations. However note that if a leak exists in the supply line then flow continuity gives **Data → Diagnostics**:

Data	$=$	**Diagnostics**	
$Q_1 - Q_2 - Q_3$	$=$	Q_{leak}	(3.44)

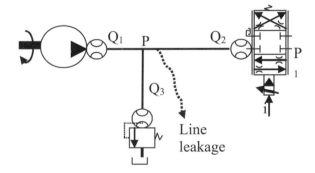

Figure 3.98. Component and line leakage detection using three flow meters

Of course, there could be measurement errors in either flow meter. If it is known that there is no leak (from inspection) and the flow measurement accuracy is $\pm\varepsilon$ then we have some simple rules:

if $Q_1 - Q_2 - Q_3 > \varepsilon$
or if $Q_1 - Q_2 - Q_3 < -\varepsilon$
and   inspection is no leak
then   fault is calibration

In essence we have produced a measure of instrument integrity. There are other ways of analysing this problem such as checking the pressure at particular operating conditions but, as in most expert systems applications, the solution used often evolves from the ingenuity of the system designer.

Example 2. A Servovalve/Motor Open – Loop Drive with Leakage

This study is based around a knowledge of the steady-state behaviour at a particular speed. It was shown in Chapter 2, for the same system, that it was possible to derive sufficiently accurate flow loss equations for the motor at a variety of speeds. These equations could be used to determine the effect of leakages in the lines and within the motor clearances, the latter by changing port plate (cross-line) "resistance" and piston "resistance" due to wear. Alternatively, as adopted here, practical tests may be carried out to determine the effect of inlet line leakage, outlet line leakage, and cross-line leakage. The approach is therefore similar to the actuator problem previously discussed and the circuit diagram is shown as Figure 3.99.

The motor is loaded via a pump/pressure relief valve with an integral solenoid-operated directional control valve. This allows a sudden load pressure to be applied to the motor to also test the dynamic behaviour of the main circuit. Leakage tests are easily achieved by connecting each motor line to tank via a restrictor valve. Motor cross-line leakage may be simulated via a connection across the motor lines, again with a restrictor valve to set the flow rate.

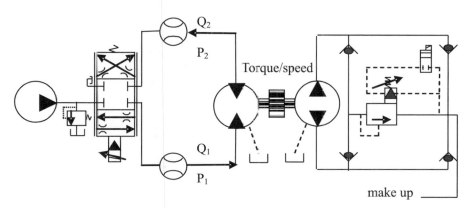

Figure 3.99. Servovalve/motor with loading circuit

Nominal leakage flows may be set to illustrate the diagnostic concept and some results are shown in Figure 3.100. It may be deduced that the "normal" flow characteristics may be represented as follows:

$$Q_{1t} = 5.5 + 0.002 \ (P_1 - P_2) \ \text{litres/min} \tag{3.45}$$

$$Q_{2t} = 4.9 + 0.002 \ (P_1 - P_2) \ \text{litres/min} \tag{3.46}$$

The development of an initial KB is based on the fact that the "normal" condition of flow rates may be approximated by a pair of simple linear equations. This is useful given the fact that the theory contains nonlinear servovalve terms as discussed in Chapter 2. In addition, each fault considered gives a unique pair of flow

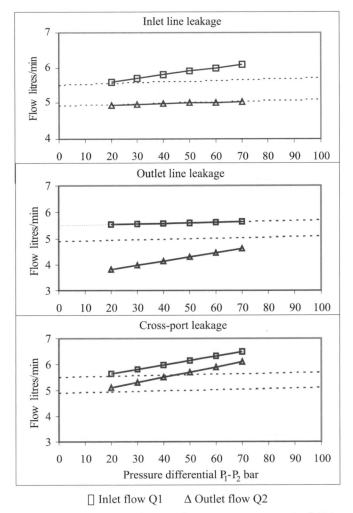

□ Inlet flow Q1   Δ Outlet flow Q2

Figure 3.100. Servovalve/motor steady-state flows at a motor speed of 500 rpm and for three leakage flow conditions

changes from the normal condition. These two aspects allow a KB to be written by using a modified theory to test in which direction the measured flow rates change.

A flow rate variation of $\pm 5\%$ is used as a measure of significant change from the normal condition. A particular feature of this approach was to test the motor speed before and after the calculations to ensure that the speed range was within 500 ± 2 rpm. The comparisons between actual flow rates and no-fault flow rates of course become more reliable as the load pressure differential increases.

A basic KB may now be developed:

fact	$q1t = 5.5 + 0.002 \times dp$
fact	$q2t = 4.9 + 0.002 \times dp$
fact	$dp = p1 - p2$
if	$q1 > q1t \times 1.05$
and	$q1 - q2 > q1t \times 1.05 - q2t$
then	leakage is inlet line leakage
if	$q1 > q1t \times 1.05$
and	$q2 > q2t \times 1.05$
and	$q1 - q2 < q1t \times 1.05 - q2t$
then	leakage is motor leakage
if	$q1 < q1t \times 1.05$
and	$q1 > q1t \times 0.95$
and	$q2 < q2t \times 0.95$
then	leakage is outlet line leakage
if	$q1 < q1t \times 1.05$
and	$q1 > q1t \times 0.95$
and	$q2 < q2t \times 1.05$
and	$q2 > q2t \times 0.95$
then	leakage is none

The subsequent KB implementation and graphics commands depend upon the expert system shell used. This application has also been significantly extended to include sensor integrity checks by using the servovalve flow equations to calculate the expected flows knowing the servovalve current and system pressures. This does result in a complex KB that, of course, relies upon sufficiently-accurate servovalve flow equations.

Example 3. Leakage Flow Detection in a Hydraulic Open-loop Lifting Circuit – the Basic KB

Consider applying flow continuity to an actuator moving at constant velocity. It is assumed that flow measurements are made at the control valve block and hence leakages possible are due to line rupture, internal leakage across the cylinder piston, and external leakage from the rod seal as shown in Figure 3.101.

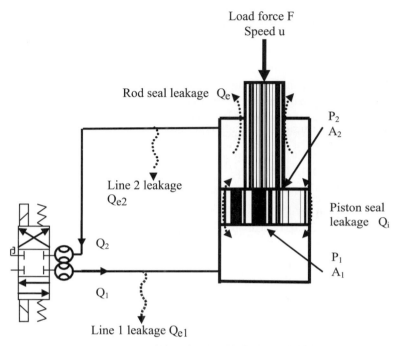

Figure 3.101. Lifting circuit with fault possibilities

An expert system is therefore required to evaluate the following eight combinations of faults:

1   no faults
2   leak from line 1
3   leak from line 2
4   internal leakage in cylinder
5   leaks from lines 1 and 2
6   leaks from line 1 and internal
7   leaks from line 2 and internal
8   leaks from lines 1, 2 and internal

Neglecting system dynamics, flow continuity for extending gives:

$$Q_1 = A_1 u + Q_i + Q_{e1} \tag{3.47}$$

$$Q_2 = A_2 u + Q_i - Q_{e2} \tag{3.48}$$

If measurements of Q_1 and Q_2 and u are available then we can rearrange the equations to:

Data on the left-hand side = **Diagnostics** on the right-hand side

$$Q_x = [Q_1 - A_1\, u] \qquad = \qquad Q_i + Q_{e1} \qquad\qquad (3.49)$$

$$Q_y = [Q_2 - A_2\, u] \qquad = \qquad Q_i - Q_{e2} - Q_e \qquad\qquad (3.50)$$

We can now determine some qualitative information from these equations, particularly on the type of leakages occurring. A simple KB program to illustrate these basic ideas would be entered as follows:

comment $\{Q_x = Q_i + Q_{e1}\}$
comment {rule 1}
if $Q_x > 0$
then   fault is leakage is line 1 and/or piston seal

comment $\{Q_y = Q_i - Q_{e2} - Q_e\}$
comment {rule 2}
if $Q_y > 0$
then   fault is leakage piston seal and/or line 2 and/or rod seal

comment $\{Q_x = Q_i + Q_{e1}\}$
comment {rule 3}
if $Q_x < 0$
then   fault is transducer line 1 and/or speed

comment $\{Q_y = Q_i - Q_{e2} - Q_e\}$
comment {rule 4}
if $Q_y < 0$
then   fault is leakage line 2 and/or rod seal and/or piston seal

We can now make a further step by noting:

$$Q_z = Q_x - Q_y = [Q_1 - A_1\, u] - [Q_2 - A_2\, u] = Q_{e1} + Q_{e2} + Q_e \qquad (3.51)$$

The right-hand side of (3.51) contains all the external leakages, and since in practice oil leaks are not desirable from a cost and safety point of view, then a warning could be initiated. Hence make this a demon:

comment {rule 5}
when $Q_z > 0$
then   fault is external leakage loss and give warning

The plant systems engineer is then able to quickly check the practical system and correct the leakage fault which should be visually obvious. Consider some further points:

- In practice we have to take into account the expected measurement accuracy so that rules will have to work within a tolerance range of $\pm\varepsilon$.
- We have only considered the lifting case. We can derive five more rules for the lowering case and improve the KB.

- This could mean a combination of rules to establish a particular fault.
- We also note that the method hinges on accurate flow rate measurements that will probably change with time during practical applications.
- Unless a reasonable constant actuator speed period exist then data comparisons will be difficult.
- This problem can be overcome by integrating the flow meter and velocity measurements.

$$Q_x = [Q_1 - A_1 \, u] = Q_i + Q_{el} \tag{3.52}$$

Integrate each term:

$$V_x = V_1 - A_1 y = V_i + V_{el} \tag{3.53}$$

V_1 = flow meter signal integrated – a standard function available, for example, with Kracht flowmeters.

y = actuator stroke used in the test so $A_1 y$ is the cylinder volume at full bore side.

V_i and V_{el} are the leakage equivalent volumes.

If we perform a full stroke test then y is known and a measurement of actuator velocity is not needed – this method is more reliable. The other flow equation is:

$$Q_y = [Q_2 - A_2 \, u] = Q_i - Q_{e2} - Q_e \tag{3.54}$$

$$V_y = V_2 - A_2 \, y = V_i - V_{e2} - V_e \tag{3.55}$$

y = actuator stroke used in the test so $A_2 y$ is the cylinder volume at the annulus side

V_i, V_{e2} and V_e are the leakage equivalent volumes.

The test procedure is therefore relatively easy to implement:

(i)   Extend and retract to full stroke.
(ii)   Compute the inlet and outlet fluid displaced either directly from the flow meter output or by finding the area under each Q_1 / time and Q_2 / time plot.
(iii) Compute the left-hand side of (3.53) and (3.55).

In this application the oil compressibility effect is negligible. There is also a degree of oil compressibility effect cancellation due to the positive change of pressure at the beginning of a stroke and a negative change of pressure at the end of a stroke. This is particularly so if line volumes dominate the total volumes on either side of the actuator. The KB for this problem now follows, the rule strategy being aided by the use of unique numerical identifiers which are easily established for this rule base size. For a much larger number of rules a binary numbering system has been found to be useful.

comment units are litres and for 1 stroke,
tol = flow meter tolerance,
a1s = bore side volume,
a2s = annulus side volume

```
fact            a1s = 4.05
fact            a2s = 2.64
fact            tol = 0.03
```

comment ************* leak rules, up stroke ***********************
```
if              v1u ≥ a1s + tol
then            equ1 = 100

if              v1u ≤ a1s − tol
then            equ1 = 200

if              v1u < a1s + tol
and             v1u > a1s − tol
then            equ1 = 300

if              v2u ≥ a2s + tol
then            equ2 = 40

if              v2u ≤ a2s − tol
then            equ2 = 50

if              v2u < a2s + tol
and             v2u > a2s − tol
then            equ2 = 60

if              v1u − v2u ≥ a1s − a2s + tol
then            equ3 = 7

if              v1u − v2u < a1s − a2s + tol
and             v1u − v2u > a1s − a2s − tol
then            equ3 = 8

fact            uoutcome = equ1 + equ2 + equ3
```

comment *************** test faults, up stroke ********************
```
if              uoutcome = 147
or if           uoutcome = 157
or if           uoutcome = 257
then            uleak is multiple
and             vup = 0
if              uoutcome = 357
then            uleak is line 2
and             vup = a2s − v2u

if              uoutcome = 167
then            uleak is line 1
and             vup = v1u − a1s

if              uoutcome = 148
or if           uoutcome = 257
```

then	uleak is internal
and	$vup = v1u - a1s$

if	$uoutcome = 368$
then	uleak is no leak
and	$vup = 0$

default	uleak is calibration drift
default	$vup = 0$

comment *************** leak rules, down stroke ******************

if	$v1d \geq a1s + tol$
then	$eqd1 = 100$

if	$v1d \leq a1s - tol$
then	$eqd1 = 200$

if	$v1d < a1s + tol$
and	$v1d > a1s - tol$
then	$eqd1 = 300$

if	$v2d \geq a2s + tol$
then	$eqd2 = 40$

if	$v2d \leq a2s - tol$
then	$eqd2 = 50$

if	$v2d < a2s + tol$
and	$v2d > a2s - tol$
then	$eqd2 = 60$

if	$v1d - v2d \leq a1s - a2s + tol$
then	$eqd3 = 7$

if	$v1d - v2d < a1s - a2s + tol$
and	$v1d - v2d > a1s - a2s - tol$
then	$eqd3 = 8$
fact	$doutcome = eqd1 + eqd2 + eqd3$

comment *************** test faults, down stroke ******************

if	$doutcome = 147$
or if	$doutcome = 247$
or if	$doutcome = 257$
then	dleak is multiple
and	$vdown = 0$

if	$doutcome = 347$
then	dleak is line 2
and	$vdown = v2d - a2s$

if	doutcome = 267
then	dleak is line 1
and	vdown = a1s − v1d

if	doutcome = 148
or if	doutcome = 258
then	dleak is internal
and	vdown = v1d − a1s

if	doutcome = 368
then	dleak is no leak
and	vdown = 0

| default | dleak is calibration drift |
| default | vdown = 0 |

The practical real-time application GUI is shown in Figure 3.102 and embodying rules for both the extending and retracting behaviour and for an actuator with a full bore total volume of 4.05 litres and an annulus total volume of 2.64 litres. The method of displaced volumes is used and Figure 3.102 shows the prediction for a particular line 2 leakage. The KB assumes a practical volume measurement accuracy of 0.03 litres, so it can be seen that the leakage is just above the threshold of detection and made possible by the use of precision flow meters that are intended for steady-state measurements only. The data shown indicate that the transient parts of the extending/retracting cycles are extremely small compared with the total stroke time.

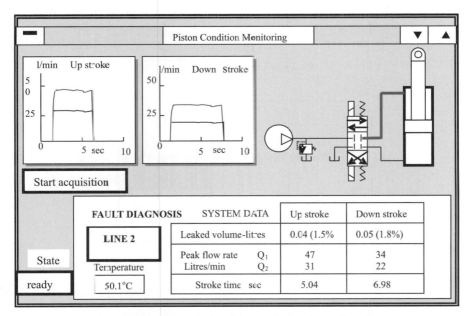

Figure 3.102. Screen copy of a cylinder drive real-time fault detection expert system [Stewart and Watton, 1992]

When using the displaced volumes method the effect of oil compressibility must be assessed. Integrating the compressible flow rate equation gives:

$$\text{Compressed volume} = \frac{V_x \, \Delta P}{\beta} \text{ at the stroke start} = -\frac{V_y \, \Delta P}{\beta} \text{ at the stroke end} \quad (3.56)$$

Where V_x and V_y are the appropriate volumes at the beginning and end of the stroke, β the fluid bulk modulus, ΔP the pressure change over the cycle concerned. These contributions will probably be small, and there is some cancellation during an on-off stroke cycle. If the circuit has lines producing similar volumes either side of the actuator then compressibility effects will almost certainly be negligible.

Example 4. A Cylinder Lifting System with Flow Control, Meter In/Meter Out, and a Check Valve Bridge, Extended KB

This example builds upon the previous example and extends the KB to include faults in a flow control valve and check valves. The previous circuit is modified as shown in Figure 3.103. The check valve bridge with integral flow control valve is used as a meter-in/meter-out unit and the design flow rate setting of 25 litres/min is used for this example. All the elements are cartridge-type resulting in a low-cost yet highly effective, accurate and compact flow control unit.

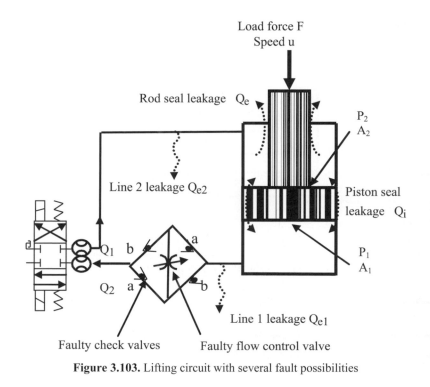

Figure 3.103. Lifting circuit with several fault possibilities

New rules are required to include the new fault possibilities and in this example a heuristic approach seemed the most direct and sensible way forward in the absence of an accurate computer simulation model of the system. Hence a further 64 practical tests are needed to obtain data that could be used to extend the KB.

A small limitation exists in that no differentiation can be made between the two check valves "a" and the two check valves "b". The additional tests may be done with appropriate check valves removed to represented the most severe "stuck-open" condition. Before this is done, it is first necessary to check how system pressures change in the presence of leakages previously discussed in Example 3.

It was found that for leakage levels appropriate for this approach, and using the displaced volume method, the presence of leakages has a negligible influence on the assumptions needed later for check valve fault diagnosis. This was certainly true for external leakages up to 10 litres/min.

With combinations of leakages and check valve faults, all the pressures and flow rates may be observed for the eight combinations of leakage fault conditions indicated in the previous Example 2. This reveals the following information:

- For the extending case, the line pressures are inherently low so the introduction of check valve faults "a" and "b" cause the line pressures to change by little more than 6 bar in most cases. This is not a useful indicator, given a pressure transducer accuracy of ± 2 bar.
- The supply pressure drops drastically by typically 40 bar for check valve "a" only, and would appear to be a useful fault indicator for all fault combinations.
- For the retracting case, the line pressures change by little more than 12 bar with the introduction of check valve faults "a" and "b". Again this measurement is not a decisive fault indicator for all the fault combinations considered.
- The supply pressure drops by typically 10 bar for all fault combinations and, unlike the extending case, is not considered to be a decisive fault indicator.

It would seem for this study that the supply pressure change emerges as the most appropriate indicator of check valve fault "a" for the extending case, and data for both extending and retracting are shown in Figure 3.104.

Conclusion 1 – Check valve "a" fault acknowledged if a 25% drop in supply pressure when extending is observed, and the desired flow control of 25 litres/min is achieved.

So what is the procedure for the check valve "b" fault condition ? Well, look at the flow rates. In this study flow rate Q_1 only is used for both extending and retracting cases and the measured data for the eight leakage fault combinations are shown in Figure 3.105.

From this data we see that a workable conclusion is:

Conclusion 2 – A fault with check valve "b" causes the retracting flow rate Q_1 to always be greater than the extending flow rate Q_1 irrespective of the condition of check valve "a".

The approach for fault diagnosis of the flow control valve is intuitive and follows from the fact that if there is a change in setting then Q_1 for both the extending and

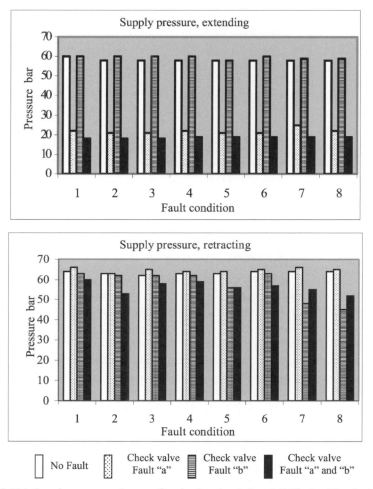

Figure 3.104. Supply pressure changes for check valve "a" and "b"faults. Nominal supply pressure is 60 bar for no-load conditions [Watton and O Lucca-Negro, 1993]

retracting cases should be the same. If a fault occurs with the flow compensator, for example being stuck, then the extending value will be greater than the retracting value due to the different pressure drops in each case. Fault diagnosis using this information does depend upon the condition of the check valves, and a measure of practical reason must be used: any faulty check valve will already have been determined by the previous method discussed. A main KB implementation structure must then be established. An additional feature that is needed is a test for the supply pressure such that there is no conflict with the individual KBs for faults. In this application a supply pressure transducer signal was used to ensure that the main pressure relief valve was functioning correctly with respect to its setting. Figure 3.106 shows the Main KB implementation structure. It will be seen that the fault priority adopted is supply pressure, lines, check valves, flow control valve.

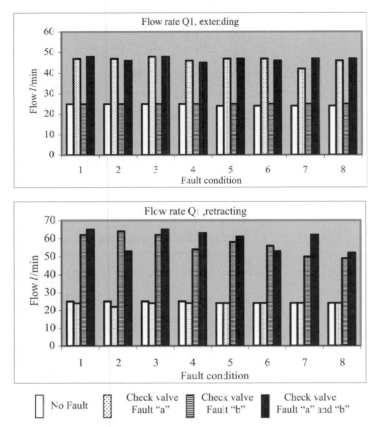

Figure 3.105. Flow rate changes in Q_1 for check valve "a" and "b"faults. Nominal supply pressure is 60 bar for no-load conditions [Watton and O Lucca-Negro,1993]

- The acquisition is started by clicking the command button. The sampling rate is calculated in order to cover the whole stroke. The stroke time is calculated from the flow rate setting and also depends on the leakages, therefore a security of 3 seconds is included.
- The graph scaling is next realised.
- Data acquisition takes place. Data processing takes place. Ten values of supply pressure are averaged before the directional valve is switched on. Steady-state pressures and flows are determined from the average centred on the middle two-thirds of actual data.
- Graphs are drawn using blocks of 50 samples to minimise noise effects and also to calculate the displaced volumes that will be used by the expert system.
- The processed data are analysed by the expert system and the conclusions are returned to the user interface. Following the conclusions evaluation, faults are displayed and highlighted on the hydraulic circuit diagram. Details of each fault condition may be obtained by clicking the appropriate command buttons.

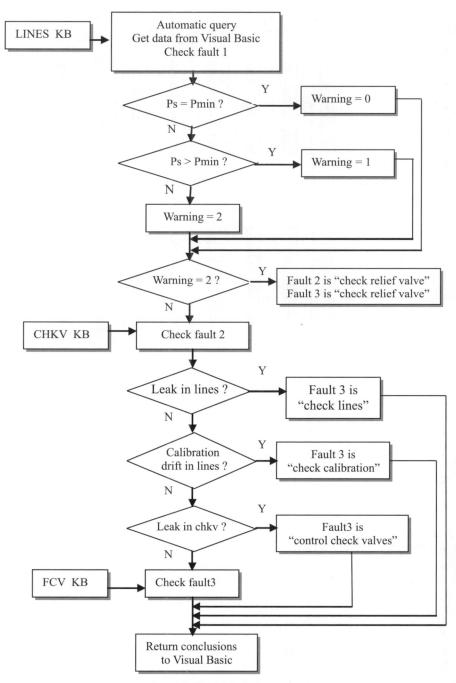

Figure 3.106. Main KB structure

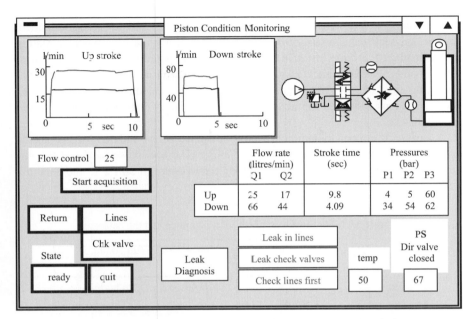

Figure 3.107. Screen display for seal leakage and check valve fault

A practical result is shown on the reconstructed GUI, Figure 3.107. A combined seal leakage and check valve fault has been set in the hydraulic circuit and the diagnosis suggests that line leakage be investigated first. This is done by clicking on the "lines" button. New data from the expert system are presented to the screen as shown in Figure 3.108.

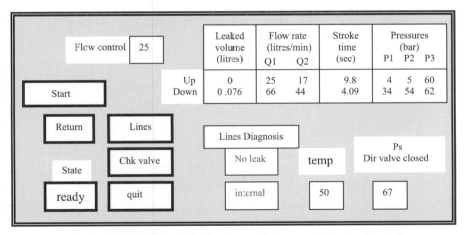

Figure 3.108. Additional display for seal leakage and check valve fault

The very low pressures when extending, due to the cylinder being unloaded, means that seal leakage cannot be detected. This is not the case when retracting and a displaced volume of 0.076 litres has been computed over a time interval of 4.09 seconds. The average leakage flow rate is therefore 1.11 litres/min. Next, operating the "check valve" button produces additional data as shown in Figure 3.109. This validates the actual circuit fault due to one of the check valves labelled as "b".

A variety of faults and combination of faults have been validated by this on-line application of expert systems concepts. Table 3.11 shows those used and the circuit-averaged steady-state data at a working temperature of $50 \pm 2°C$ and a supply pressure of 65 bar.

Table 3.11. Some fault combinations examined and the data sets

Test	Q1 (l/min)	Q_2 (l/min)	P_1 (bar)	P_2 (bar)	time (sec)	Fault condition
1 Ext	25	17	4	5	9,60	No faults
Ret	25	17	40	61	9,79	
2 Ext	25	17	4	6	9,77	Line 1 leak
Ret	25	17	40	62	9,74	
3 Ext	25	16	4	5	9,78	Line 2 leak
Ret	25	17	41	62	9,88	
4 Ext	25	15	4	5	10,48	Multiple leaks
Ret	25	19	41	61	8,76	lines 1 and 2
5 Ext	25	15	5	4	11,29	Multiple leaks
Ret	19	15	42	63	11,67	flow c v, lines 1 and 2
6 Ext	25	15	5	3	10,25	Line 2 leak
Ret	20	15	42	62	13,04	flow c v, line 2
7 Ext	25	16	5	4	9,82	Flow control valve
Ret	19	13	42	63	12,66	
8 Ext	49	32	6	9	5,17	Check valve "a"
Ret	25	17	39	62	9,78	
9 Ext	25	17	4	5	9,8	Piston seal fault check
Ret	66	44	34	54	4,09	valve"b" fault
10 Ext	23	15	3	4	10,24	Low system pressure
Ret	22	15	30	37	10,86	piston seal fault
11 Ext	23	15	4	5	10,46	Flow control valve
Ret	23	16	40	62	10,46	setting decreased

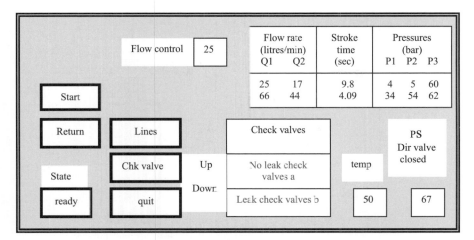

Figure 3.109. Additional display for seal leakage and check valve fault

Example 5 Position Control of a Servovalve/Cylinder Drive

The displaced volume method for leakages equally applies to closed-loop control systems and in this study step response tests were able to detect leakages to a minimum value of 0.14 litres/min.

A particular feature of closed-loop systems is that the existence of the various leakages previously discussed results in a position steady-state error, the sign and magnitude of which depends upon the particular leakage fault. This aspect is convenient as a stand-alone KB since it does not require expensive flow meters in practice, just measurement of the servovalve error current or equivalent position transducer error voltage. Alternatively, if flow metering is to be used then the steady-state error phenomenon may be added to the KB to give more confidence to the fault prediction. Consider therefore the position control system shown in Figure 3.110.

Position was measured using a linear variable displacement transducer and a variable load force was created using an additional load single-rod actuator with flow make-up. A series of extending and retracting tests then produced steady-state error data. Inherent, although small, steady-state errors exist due to the practical performance of servovalve drives. In practice this effect is usually designed to an acceptable level by both servovalve choice and gain/control compensation selection. However steady-state errors due to leakage cannot be removed since the servovalve spool must be open to service the leakage. Some typical measurements of error are shown in Figure 3.111 and for extremely low values of leakage flow rate. The position transducer gain = 0.023 mm/mV.

In practice a tolerance must be included in the extended KB due to the steady-state error with no fault, but the KB extension is relatively easy as deduced from Figure 3.111 and shown in Table 3.12.

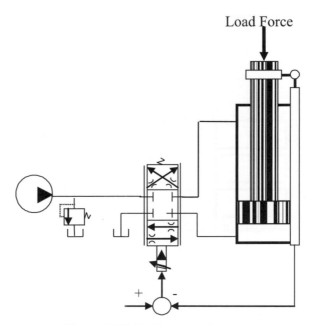

Figure 3.110. Position control system

It is then a simple matter to add new rules to the KB to check whether or not the changes indicated in Table 3.12 have occurred. This aspect is included in Chapter 3 where the full study is completed by the addition of a third approach using artificial neural networks.

Table 3.12. Steady-state error changes with various leakages present

Fault	Error extending mV	Error retracting mV
none	$-17 \rightarrow -20$	$-7 \rightarrow -11$
Line 1	increase	increase
Line 2	decrease	decrease
Internal	increase	decrease

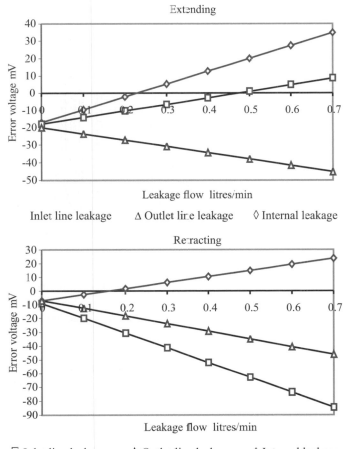

Figure 3.111. Steady-state position error voltages for a servovalve/cylinder drive under various leakage conditions.

3.10 Object-oriented Expert Systems Approach for Multiple Subsystems, and its Application to a Hot Steel Strip Rolling Mill

The hot strip finishing mill being considered was discussed earlier with respect to the data acquisition and monitoring strategy, and a schematic is shown as Figure 3.112.

The main objective was to successfully introduce a robust on-line fault detection system for the first of the three major hydraulic control systems, the work roll bending system. A particular feature of the advanced system is the interconnection of the data acquisition and data processing computer with the Gensym G2 Expert System computer for the graphics presentation of the mill, and actions required as a result of the rule base decision process. When considering the specific collaboration with Corus plc, Port Talbot UK, a clinical examination of relevant signals in

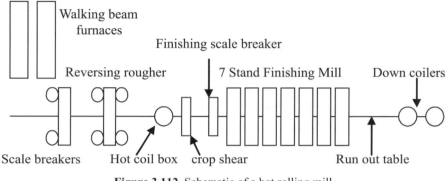

Figure 3.112. Schematic of a hot rolling mill

conjunction with the dynamic performance of the mill stands was required. Fault data from the condition monitoring Expert PC is then distributed through the present network protocol. Within the on-line system, faults are generally specified as being mechanical, electrical or hydraulic, and are available within the networked system for clear recognition by the user. The networked computer architecture is shown as Figure 3.113.

When considering such large applications as being discussed here, it is preferable to streamline the real-time intelligent monitoring and fault diagnosis strategy. An object oriented approach is particularly useful when a complex system has a number of similar hydraulic components or subsystems. The approach is to represent components as "objects" organised in a hierarchical class structure. An

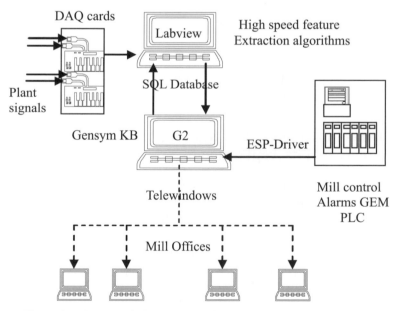

Figure 3.113. Networked computer architecture [Courtesy of Corus plc]

object inherits properties from multiple object classes and any object or group of objects can be reproduced, each reproduction also inheriting all the properties of the original object. Objects, rules, and procedures can be grouped into library modules that may be shared with new applications thus streamlining the development process. They are autonomous in the sense that they operate when activated by an external message or another object. Many engineering systems have multiple objects with common attributes. Such a group is known as a class and each group belong to the group is called an instance. A class at a higher level has a smaller number of common attributes than at a lower level, and attributes at a higher level are passed to the lower levels.

Consider one steel strip rolling mill stand as shown schematically in Figure 3.114, and representing just one typical stand.

Each of the four WRB cylinders and the two AGC capsules are controlled by a servovalve and most of the control systems are reproduced on the other mill stands. Clearly, expert system monitoring design can be significantly aided by the object-oriented approach using task-specific objects rather than lines of repeated computer code.

Consider, for example, the object named "valve block" shown in Figure 3.115, and representing initially just one block for WRB control of one stand. It is designated as the highest-level object in the class structure of the expert system and contains descriptions and configurations for all the low-level objects.

The workspace named "system 12 valve block classes" has been designed so that class-specific attributes (for example common features of the hydraulic components) are defined only once at the highest level. All the lower level sub-classes (such as a servovalve) inherit attributes and values by association from the higher

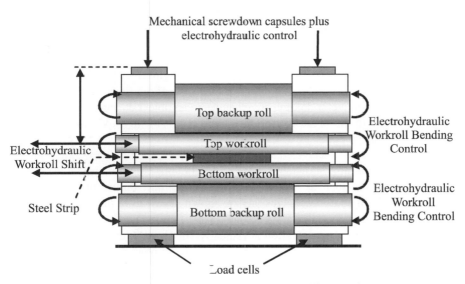

Figure 3.114. Schematic of a steel strip rolling stand

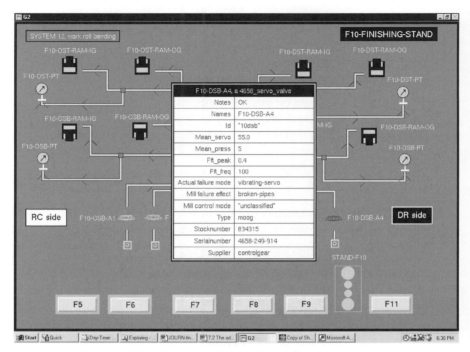

Figure 3.115. Inherited attributes of the FMEA for a servovalve

level "valve block" object. WRB utilises four cylinders, drive side top (dst), drive side bottom (dsb), operator side top (ost), operator side bottom (osb).

It will be seen that four classes of ram have therefore been developed. This also aids the hydraulic mimic design since it is then similar to the hydraulic circuit drawings supplied by the manufacturer. Two icons are shown for the two pressure transducers used for WRB pressure control, and representing top and bottom cylinders. The advantage of using the G2 expert system is that once the classes are created then an instance (identical copy of the original) may be generated any number of times to build the user interfaces for each of the seven stands. Using the hierarchical design methodology, any changes to an object during the development stage are made to the highest level object and automatically updated to every instance of the object class.

Once the instance receives data from the database, then the default colour becomes green. The table shown in Figure 3.115 contains common inherited descriptions of attributes such as "actual failure mode", "mill failure mode"and "mill control mode". In addition conclusions are shown drawn from the KB, in this case a faulty servovalve due to internal erosion of the flapper/nozzle assembly causing excessive vibration. This vibration at typically 100 Hz at a level of 0.4 V rms has been investigated in some detail and shown to cause pipework failures with subsequent fluid losses. Note that in this application there are 28 servovalves. Examples of some of the KB rules and the method of description follow. The main objective

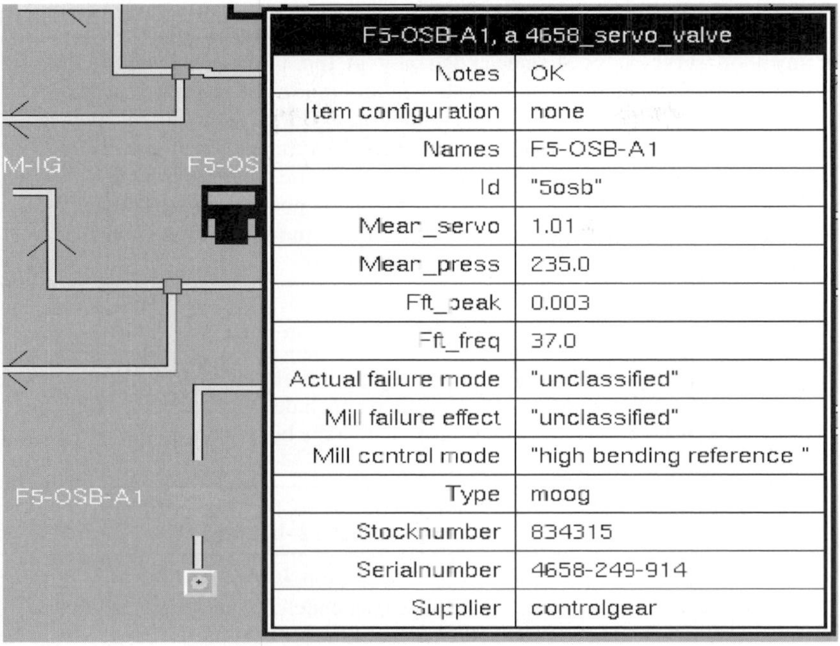

F5-OSB-A1, a 4658_servo_valve	
Notes	OK
Item configuration	none
Names	F5-OSB-A1
Id	"5osb"
Mear_servo	1.01
Mear_press	235.0
Fft_peak	0.003
Fft_freq	37.0
Actual failure mode	"unclassified"
Mill failure effect	"unclassified"
Mill control mode	"high bending reference "
Type	moog
Stocknumber	834315
Serialnumber	4658-249-914
Supplier	controlgear

for any 4658_servo_valve S connected to any pt_top
if the mean_press of S > 50 then conclude that the
mill-control-mode of S = "high bending reference"

Figure 3.116. Table for a no-fault condition at a high bending reference

is to inform the operator of the type and severity of the hydraulic control fault, and a further feature is the use of different colours (red, amber, green) to indicate the alarm status.

The move towards a satisfactory KB for this application by definition requires the pooling of many resources including mill designers, operators, maintenance personnel, external consultants, management personnel. The development being discussed here calls upon vibration frequency information, via servovalve current loop data and pressure transducer data, obtained from the database of information analysed via the National Instruments LabView data acquisition system and associated software. Considering work roll bending, faults are identified during the different control modes applicable to the rolling strategy adopted. Figure 3.116, for one stand, indicates a mean pressure of 235 bar, which is identified as "high bending reference".

It is concluded that the servovalve mean current of 1.01 mA and the dominant frequency component of 37 Hz, with a magnitude of 0.003 V, is not significant and does not require fault classification.

Set control-offset > 5, -bottom

for any 4658_servo_valve S connected to any pt_top if the mean_servo
of S >= 5 then conclude that the actual-failure-mode of S = "large-current-
offset" and conclude that the mill-failure-effect of S = "unclassified" and
change the body icon-colour of S to red and show the message-board and
inform the operator for the next 5 minutes that " The control circuit
incorporating [the name of S] is subject to a large positive current offset [the
mean_ servo of S] Hz. Please inve stigate [the name of S] WRB control at the
next roll change—Mr X Y Z (Fluid Power Group)"

Set control-offset >2, <5 -bottom

for any 4658_servo_valve S connected to any pt_top if the mean_servo
of S < 5 and the mean_servo of S >= 2 then conclude that the actual-failure-
mode of S = "small-current-offset" and change the body icon-colour of S to
orange

Reset FMEA if mean servo >-2 and <2 -bottom

for any 4658_servo_valve S connected to any pt_top if the mean_servo
of S >= 2 and the mea n_servo of S < -2 then conclude that the actual-failure-
mode of S = "unclassified" and conclude that the mill-failure-effect of S =
"unclassified" and change the body icon-colour of S to green

Figure 3.117. FMEA rules for servovalve current offsets

In addition to servovalve vibration, there are other fault conditions that manifest
themselves as positive or negative current offsets. High values of offset can cause
the WRB control system to trip, so the icon colour of the servovalve is changed to
red. The high-offset current value selected here is 5 mA and an appropriate FMEA
set of rules, using the KB format, is shown in Figure 3.117. Some detailed thought
has to be given to the actual cause of offset since it could result from either
a leakage condition, servovalve deterioration or spool underlap, or changes in
system friction levels. The approach previously outlined serves merely to inform
the operator that offsets are changing. This is yet another example that complex
KBs can easily evolve from an ostensibly simple observable problem. The use of
offset classification must also be adjusted as experience is gained to minimise the
appearance of "orange" warning messages which actually do not represent
a serious effect on the rolling performance and hence strip quality.

An example of a serious hydraulic failure mode is illustrated in the FMEA table
shown in Figure 3.118 together with the corresponding FMEA rule. In this
example a vibrating servovalve is detected during WRB balance mode (low
pressure setting) and resulting in a broken pipe.

Pressure transducer faults due to either failure or calibration changes require
immediate attention otherwise the control circuit may completely fail. Calibration
faults, for example failing to indicate pressures below 6 bar, result in the inability

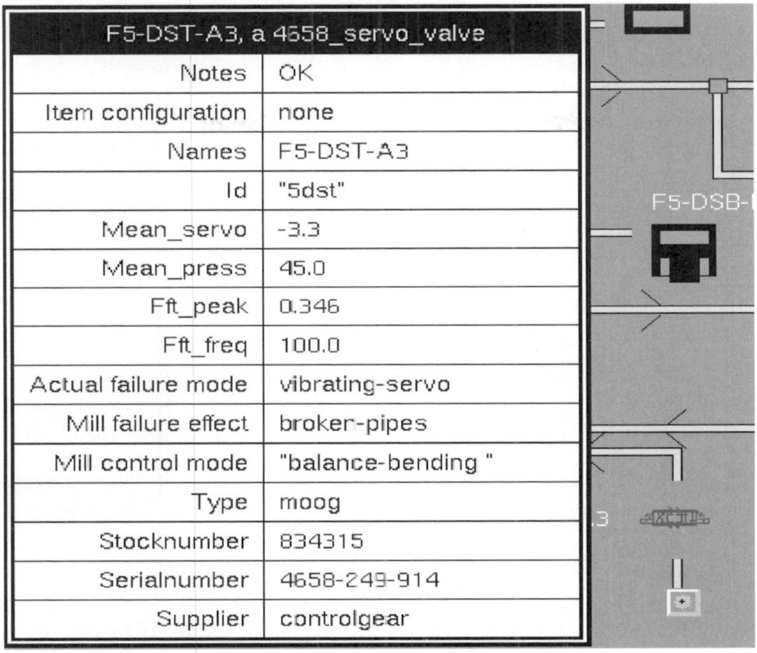

F5-DST-A3, a 4658_servo_valve	
Notes	OK
Item configuration	none
Names	F5-DST-A3
Id	"5dst"
Mean_servo	-3.3
Mean_press	45.0
Fft_peak	0.346
Fft_freq	100.0
Actual failure mode	vibrating-servo
Mill failure effect	broker-pipes
Mill control mode	"balance-bending "
Type	moog
Stocknumber	834315
Serialnumber	4658-249-914
Supplier	controlgear

for any 4658_servo_valve S if the fft_freq of S >= 80 and the fft_freq of S <= 110 and the fft_peak of S >= 0.3 then conclude that the actual - failure-mode of S is vibrating-servo and conclude that the mill-failure-effect of S is broken pipes and change the body icon colour of S to red and show message-board and inform the operator for the next 10 minutes that "The [the name of S] has been vibrating at = [the fft_freq of S] Hz and a vibration level = [the fft_peak of S] V rms. Please remove and replace with a [the type of S] 4658_servo_valve using the stock number [the stock number of S] available from the   supplier [the supplier of S]—Mr X Y Z (Fluid Power Group)"

Figure 3.118. FMEA table and rule for a serious servovalve fault condition

to reach low bending reference. The control error increases as the WRB system attempts to lower the pressure and as a result causes the 'mean servo' value to increase. The likelihood is therefore high that a large positive error is associated with the pressure transducer. An FMEA table and associated rule is shown in Figure 3.119.

From these example it can been seen how the comprehensive KB has been built up. In this application the expert system has been configured to fire FMEA rules at designated scan intervals of 60 seconds and in a ranked priority. FMEA rules for

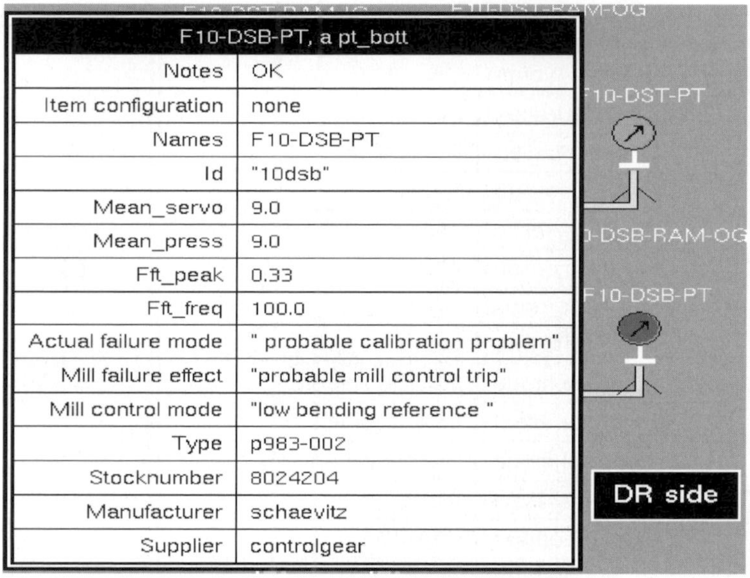

F10-DSB-PT, a pt_bott	
Notes	OK
Item configuration	none
Names	F10-DSB-PT
Id	"10dsb"
Mean_servo	9.0
Mean_press	9.0
Fft_peak	0.33
Fft_freq	100.0
Actual failure mode	" probable calibration problem"
Mill failure effect	"probable mill control trip"
Mill control mode	"low bending reference "
Type	p983-002
Stocknumber	8024204
Manufacturer	schaevitz
Supplier	controlgear

for any pt_bott P connected to any 4658_servo_valve if the mean_press of P < 10 and the mean_servo of P >= 9 then conclude that the actual-failure-mode of P = "probable calibration problem" and conclude that the mill-failure-effect of P = "probable mill control trip" a   nd change the face icon-colour of P to red and show message-board and inform the operator for the next 20 minutes that "The pressure transducer[the name of P] may not be achieving a low bending reference and may need to be replaced at the next down shift"

Figure 3.119. FMEA table and rule for a pressure transducer failure at low bending reference

the servovalves, pressure transducers and rams have been configured to allow the most dominant fault to be shown for the majority of the time. In particular a 'vibrating servo' fault is shown for 80% of the time and the 'small negative offset' fault for the remaining time. The inference engine of the expert system examines the 'scan interval' and the 'rule priority' of the attributes in each of the FMEA tables, for example the table in Figure 3.118 is given a priority of 1. The system consists of seven connected screen mimics for each mill stand. The FMEA tables are normally hidden and the message-board presents the appropriate information containing a predictive maintenance proposal. The user may "double-click" on any of the hydraulic components to obtain the current FMEA table and the screens are updated every time the steel strip passes through the mill, typically every three minutes. Figure 3.120 shows the final design.

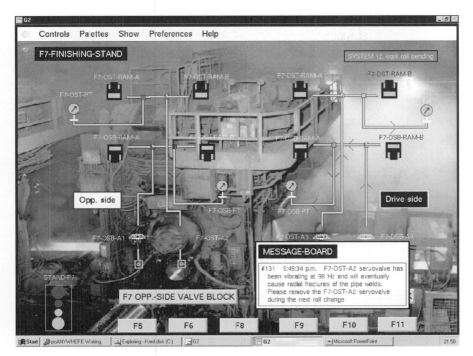

Figure 3.120. Design of the mill monitoring and diagnostic expert system

The buttons shown at the bottom of the screen mimic allow the user to seamlessly navigate between the seven mill stands, the third stand being interrogated. The expert system has been programmed to provide a prediction condition monitoring capability for mechanical, electrical and system faults including probable fluid leakage.

Chapter 4

Common Faults and Breakdowns that can occur in a Hydraulic Circuit

The following information is taken directly from the publication "Analysis and Explanations of the Causes of Destruction of Pumps in Service" by Hagglunds Denison Ltd and "Vane Troubleshooting Guide" by Denison Hydraulics Ltd. These publications contain useful textual information and many colour photographs of faults that commonly occur in vane and piston pumps. The list of problems is not exhaustive and represents general product failures throughout the whole fluid power industry and not specifically one production unit. This list is therefore based upon experience gained over many man-years, and thanks are extended to Denison Hydraulics UK for support and permission to reproduce just part of the large amount of information available from them.

4.1 Pumps and Motors

FAILURE	CAUSE	SOLUTION
1. Pumps and motors		
1-1 Insufficient or no output.	a) Incorrect pump rotation, coupling key loose.	a) Correct electrical connections. Check that the key is in place.
	b) Oil level too low.	b) Top up reservoir.
	c) Clogged strainer or too small.	c) Clean or replace.
	d) Incorrect pressure control setting.	d) Adjust setting.
	e) Suction gate valve closed, or partially closed	e) Adjust to open or replace.
	f) Air bubbles in the circuit.	f) Bleed the circuit of air and allow air to settle out of oil in reservoir.

FAILURE	CAUSE	SOLUTION
	g) Suction joints allowing air into system.	g) Change seals or replace.
	h) Broken pump shaft.	h) Find the cause (overloading or poor alignment of the pump): change the shaft.
	i) Incorrect fluid for application.	i) Check system specification reference to H.F. Fluids. Replace fluid.
	j) Oil too cold (viscosity too high).	j) Heat oil to recommended temperature.
	k) Pump not primed.	k) No air bleeder device on the discharge (Loosen one connector on the pipe & bleed).
	l) Suction height excessive.	l) Reduce the height. Check pump specification, particularly for H.F. Fluids.
	m) Pump's rotation speed too high.	m) Reduce rotation speed to that recommended.
	n) Sealed reservoir (air-tight tank) or breather blocked.	n) Install new breather or an air filter with sufficient capacity.
1-2 Noisy Pump or Motor	a) Cavitation	a) Prime the pump's circuit (also refer to 1-1 j and 10-2) adjust or control the hydraulic motor's deceleration.
	b) Entry of air via the intake line joints.	b) Change the connector or seals and check air-tightness of the pipe.
	c) Shaft seal allowing entry of air.	c) Renew seal. Check the shaft alignment.
	d) Aeration	d) Bleed the circuit (see also 10-2).
	e) Air intake system in reservoir clogged or missing.	e) Clean or install air intake system.
	f) Strainer is too small.	f) Install a larger strainer or clean it.
	g) Diameter of pump intake tubes too small or the pipe is blocked.	g) Replace with larger diameter pipe. Remove blockage. Check fluid velocity.
	h) Air intake at pump housing.	h) First check housing bolts are at correct torque, and condition of the seals compatible with fluid.
	i) Broken vane spring.	i) Change cartridge. If premature, investigate cause.
	k) Worn or damage in the pump motor.	k) If premature, investigate cause. Repair or replace.

FAILURE	CAUSE	SOLUTION
	l) Pump or motor subjected to strain on mounting.	l) Check alignment of the mounting and tighten the nuts evenly.
	m) Contamination in the supply circuit.	m) Remove the contaminants and if necessary clean the circuit.
	n) Contaminated circuit.	n) Drain system, flush and fill to NAS/ISO standard. Analyse particles to determine their type and their particle size.
	o) Flattened intake suction pipe.	o) Replace
	p) Oil temperature too high.	p) Check the circuit design to establish the cause (cooler/) see also 1-3. System heat extraction capacity must match power losses.
	q) Failure in the boost pump.	q) Find the reason and repair boost pump.
	r) Noisy tank (Resonance).	r) Change the tank's position or attachment, install an anti-noise system. Re-site pump.
	s) Porous intake flexible hose.	s) Change the flexible hose.
	t) Vibrations in the circuit. (Component resonance.)	t) Find the cause and repair (see also 12-1).
	u) Other pump or motor defects.	u) If none of the solutions referred to herein, return to the factory.
	v) Oil level too low.	v) Top-up reservoir. Check for system leaks.
	w) Incorrect functioning of the intake valve.	w) Repair or remove this valve.
	x) Suspect an incorrect fluid.	x) Replace with reference to H.F. Standards.
	y) Rotation speed too high.	y) Adjust to recommended speed.
1-3 Pump or Motor giving off too much heat.	a) Suspect an incorrect fluid.	a) Replace with reference to H.F. Standards.
	b) Excessive fluid velocity in the circuit.	b) Revise pipe size or derate flow.
	c) Oil level too low.	c) To up reservoir.
	d) Pump or motor's rotary assembly worn.	d) Renew cartridge. If premature, investigate cause.
	e) Radial or axial load too high.	e) Check the alignment and radial load limit to pump specification.

FAILURE	CAUSE	SOLUTION
	f) Increase in the initial speed by modification of the circuit.	f) Check the maximum pressure; if necessary change the type of pump (flow to high) and install correspondingly sized pipes.
	g) Poorly sized cooler.	g) Increase the capacity of the cooler. (See also 14-1)
	h) Blocked or contaminated cooler.	h) Find the cause and repair (deposits, tartar, bacteria in the water circuit).
	i) Too little difference between the set maximum, and the operating pressure.	i) Increase the set maximum pressure or reduce the operating pressure.
	k) Relief valve set too high.	k) Adjust to correct setting.
	l) Error in the choice of the relief valve.	l) Replace with one to meet specification.
	m) Relief valve malfunction.	m) Repair or replace.
	n) Error in the internal seals.	n) Replace with reference to H.F. Standards.
	o) Filter – clogged or too small causing excessive pressure drop.	o) Clean the filter or replace the cartridge, change the filter.
	p) Rotation speed too high.	p) Reduce to recommended pump specification speed.
	q) Cavitation	q) Check the pump priming and bleed the circuit. (See also 1-1j and 10-2)
	r) Aeration	r) Bleed the circuit. (See also 10-2.)
	s) Air breather clogged.	s) Clean the system. (See also 1.1m).
	t) Contaminated circuit.	t) Drain system, flush and fill to NAS/ISO standard, and if necessary scour the pipes and clean them again.
	u) Flattened inlet pipe.	u) Replace the inlet pipe.
	v) Boost pump failure.	v) Determine the cause and repair boost pump.
	w) Other defects in the pump or motor.	w) If none of the remedies referred to herein, return to factory.
1-4 Lack of pressure from the pump or system.	a) Incorrect pressure setting.	a) Locate valve and adjust setting.
	b) Relief valve malfunction.	b) Locate unit, clean, adjust, repair or replace.

FAILURE	CAUSE	SOLUTION
	c) Faulty electrical circuit (solenoids).	c) Check the electrical circuit and signal to appropriate solenoids.
	d) Leak in the circuit (cylinders).	d) Check seals and replace any faulty ones.
	e) Error in the circuit assembly.	e) Check the pipework and rectify accordingly.
	f) Broken pump shaft or shaft key missing.	f) Find the cause (pump subjected to strain?). Change the shaft, put the key back in place.
	g) Incorrect setting of the by-pass contacts.	g) Modify the setting of the contacts. (Pressure switches)
	h) No flow from pump.	h) See 1-1.
	i) Poor fluid quality.	i) Drain system, flush and fill to NAS/ISO standard. H.F. Specification
	k) Faulty drive mechanism belt.	k) Repair mechanism. (Find the cause.)
	l) Slipping of the drive mechanism belt.	l) Adjust or replace the belt.
	m) Contaminated circuit.	m) Drain system, flush and fill to NAS/ISO standard.
	n) Poor seals and coverings, suspect fluid compatibility.	n) Replace with recommended parts with reference to H.F. Specification.
1-5 Loss of motor speed.	a) Supply pressure too low.	a) Increase this pressure.
	b) Back pressure too high.	b) Check the circuit.
	c) Front plate not making contact. (Port plate).	c) Strip vane motor and repair.
	d) Worn or damaged motor (wear by abrasive particles).	d) Replace or repair. If premature investigate cause.
	e) Oil temperature too high.	e) Check the circuit (cooler). See 1-3.
1-6 Poor speed control.	a) Excessive system leakage relating to fluid temperature change.	a) Identify critical unit (s), repair, replace or incorporate compensated flow.
1-7 The motor does not work.	a) Torque too low.	a) Check–increase the initial pressure if necessary.
	b) Large internal or drain leak.	b) Check the internal parts. Deterioration by contamination or cavitation. Repair or replace.
	c) Faulty O rings on the front plate	c) Replace these O rings and check that the front plate runs freely. (Check the axial balance).

FAILURE	CAUSE	SOLUTION
	d) Suspect pump flow.	d) Repair or replace the pump with one which has a greater flow. (Check input power)
	e) Motor too small. (Displacement).	e) Replace it with a larger model. (See slower speed)
1-8 Too much play in the shaft.	a) Defective bearing.	a) Replace the bearing.
	b) Too high a radial or axial load.	b) Ensure that the maximum authorised load is not exceeded. Install lot-shaft if necessary.
	c) Badly balanced coupling. (Eccentric throw)	c) Balance or change the coupling.
1-9 Leak in the pump or motor.	a) Insufficient sealing on connections.	a) Check the seals.
	b) Poor sealing of the shaft seal.	b) Look for the cause, change the seal.
	c) Leaks in the housing.	c) Check for porosity and if necessary change the housing.
	d) Damaged mating faces.	d) Remove imperfections on the faces or preferably send the pump or motor for rectification or replacement.
	e) No deceleration valve on the motor circuit (braking pressure too high = destruction of the motor).	e) Install a declaration valve and feed valves. (Counterbalance and check)

4.2 Directional Valves

FAILURE	CAUSE	SOLUTION
2. Directional Valves.		
2-1 Sticking spools.	a) Sticking spool due to distortion.	a) Loosen fixing screws, evenly tighten. (Check flatness of the contact areas)
	b) Particles in the circuit.	b) Drain system, flush and fill to NAS/ISO standard.
	c) Poor fluid quality.	c) Drain system, flush and fill with ref to H.F standards, NAS/ISO standard.
	d) Water in the circuit.	d) Check cooling system with ref to temp cycle. Reservoir condensation (see 14-2)

FAILURE	CAUSE	SOLUTION
	e) Thickening of the anti-corrosion oil (as a result of being stored too long).	e) Clean the spool and if necessary circulate the oil to remove deposits.
	f) Error in valve assembling.	f) Carefully follow the sectioned diagram in the service literature.
	g) Wrong connectors.	g) Replace them with the right ones.
	h) Oil pump too high.	h) See 10-3 and 14-1.
	i) Oil speed too high, loss on load and unbalancing of the spool Bernoulli forces.	i) Install a valve with a larger capacity.
	k) Piping subjected to strain (longitudinal extension) causing valve distortion.	k) Piping too short or establish compensation curves only for large differences in temp and pressure.
	l) Oil too cold.	l) Heat system to recommended temperature.
	m) Faulty spool.	m) Repair or replace spool. (Be careful of the state of the bore in the body)
	n) Drain line under pressure or missing. (On externally drained valves.)	n) Establish an independent direct drain line to the tank.
2-2 The solenoid does not work.	a) Burnt out solenoid.	a) Check, replace sol
	b) Blocked spool.	b) See 2-1
	c) No current at solenoid.	c) Check the cables and fuses for power supply to solenoid.
	d) Fault in the electrical circuit.	d) Check circuit. The solenoid may burn out for the following reasons: under/over voltage, two solenoids energized at the same time, unbalancing of the spool, back pressure in the drain, temp too high in the = change in the resistance valve and therefore increased consumption = increased heating = distortion of the nylon core support and guide.
2-3 The two-stage distributor does not work.	a) Blocked spool.	a) See 2-1
	b) No pressure.	b) Check the circuit (see 1-4/5-2 & 5-3).
	c) No pilot line.	c) Install this line (See 2-1n).
	d) Blocked pilot line.	d) Clean line (passage build up of silt?). Check for a blocked orifice.

FAILURE	CAUSE	SOLUTION
	e) The slide does not go back.	e) Choke block is not set correctly, check the diaphragm screw in the electric pilot is not blocked.
2-4 The valve gives off too much heat.	a) Temperature in the circuit too high.	a) Check circuit and cooling system.
	b) Poor fluid quality.	b) Follow the Manufacturer's instructions. Drain system, flush and fill with reference H.F. Standards and filtration NAS/ISO standard.
	c) Dirty circuit.	c) Drain system, flush and fill to NAS/ISO standard
	d) Fault in the electrical circuit.	d) Check the electrical circuit.
	e) Blocked spool.	e) See 2-1.
	f) Faulty spool. (Jamming by particles)	f) Service or replace spool.
2-5 Noisy valve.	a) valve is too small (high fluid velocity).	a) Increase valve and pipe size.
	b) Vibrations in the circuit.	b) Tighten the piping (see 12-1)
	c) No choke pack.	c) Install a choke pack.
	d) Spool incompatible with the circuit.	d) Change to an appropriate spool.
	e) Blocked spool.	e) Check to see if the circuit is dirty. (See also 2-1 n and NOTE)
2-6 Leak in the valve.	a) Poor connection seals.	a) Check the seals (special fluids and compatible seals).
	b) Defective or badly assembled seal.	b) Change the seal and lubricate it.
	c) Sealed parts not tightened to correct setting.	c) Tighten to required torque specification.
	d) Fault in the valve or spool.	d) Crack in the body? Change the valve or spool.
	e) Drain line under pressure.	e) Install an independent drain line direct to tank.
	f) No drain line.	f) Install the drain line direct to tank.

4.3 Servovalves

FAILURE	CAUSE	SOLUTION
3. Servovalves		
3-1 Blocked For a worn servoval-ve, send it to the specialist department for repair.	a) Dirty supply tube.	a) Check cleanliness, if not, strip, clean, flush etc.)
	b) Clogged filter.	b) Check the cleanliness of the circuit and clean the filter or replace the cartridge element.
	c) Mechanical return of the slide. Blocked feedback.	c) Establish the cause (valve under strain) change the valve.
	d) Valve under strain (distorted body).	d) Loosen the valve and then tighten evenly.
	e) Other causes. (See 2-1 b, c and d.)	e) See 2-1 b, c and d.
3-2 The ser-vovalve does not work.	a) Fault in the electrical circuit.	a) Check the circuit and ampli-fier.
	b) Damaged magnetic circuit.	b) Repair or replace the circuit.
	c) No differential current	c) Check the electrical circuit.
	d) No pressure	d) Check the circuit. (See 1-4/5 and 5-3)
	e) Dirty supply tube.	e) See 3-1 a.
	f) Clogged filter.	f) See 3-1 b.
	g) Mechanical return of the slide.	g) See 3-1 c.
	h) Valve under strain.	h) See 3-1 d.
	i) Poor fluid quality	i) Follow the instructions (for special fluids consult manu-facturer)
	k) Oil too thick.	k) Flush the valve, if necessary change it, replace the oil.
	l) Oil temperature too high.	l) Reduce the initial pressure or increase the coolant.
	m) Capacity of the valve too small.	m) Install a valve with larger capacity.
	n) Faulty valve.	n) Change the valve. (See note)
3-3 The ser-vovalve gives off too much heat.	a) Error in the current.	a) Check the current and the electronic base plate, if nec-essary modify it.
	b) Mechanical return of the slide, feedback blocked.	b) See 3-1 c.
	c) Other causes.	c) See 2-4 c and d.
	d) Other causes.	d) See 2-6 a and c.

4.4 Check Valves

FAILURE	CAUSE	SOLUTION
4. Check valves		
4-1 Blocked check valve	a) Spool tight, not free.	a) Loosen the screws and then tighten them evenly.
	b) Assembly error.	b) Follow the assembly instructions.
	c) Check valve seat out of line.	c) Install a new seat and check that it is positioned correctly.
	d) Tight pilot piston.	d) Find the cause and repair or change the pilot piston.
	e) Drain line omitted.	e) Install drain line direct to the tank.
	f) Drain line pressurised.	f) Install a separate drain line direct to the tank.
	g) Other causes.	g) See 2-1 b and f.
4-2 Leak in the valve.	a) Faulty check valve.	a) Repair or replace the check value and check the cleanliness of the circuit.
	b) External leaks.	b) Replace seals, refer to compatible materials, HF standards.
	c) Loose assembly.	c) Rectify.
4-3 Resonance in the valve.	a) Timing block missing.	a) Install this circuit. (Stronger diaphragm screw or spring.)

4.5 Pressure Limiters

FAILURE	CAUSE	SOLUTION
5. Pressure limiters.		
5-1 Cavitation in the pressure limiter	a) Faulty seat.	a) Replace part.
	b) Faulty pilot cap.	b) Repair/replace.
	c) Oil speed too high.	c) Install limiter with bigger capacity.
	d) Poor quality oil.	d) Follow the instructions. For special fluids consult manufacturer check HF standards.
	e) Dirty circuit.	e) Drain system flush and fill respecting NAS/ISO filtration standard. Examine particles.
	f) Pulsating limiter.	f) Foaming/polluted fluid. Decant/bleed. If necessary change the fluid.

FAILURE	CAUSE	SOLUTION
5-2 Pressure limiter blocked.	a) Limiter under pressure (tight spool).	a) Loosen the nuts and then tighten them evenly.
	b) Oil temperature too low.	b) Operate the pump at low pressure or install a heating system. Heat to recommended temp.
	c) Piping under stress (valve distortion).	c) Install elbows to compensate (only if there are very big differences in temperature).
	d) Drain line under pressure or missing.	d) Install a drain line or separate it from the tank return.
	e) Other causes. (See 2-1 b and h)	e) See 2-1 b and h.
5-3 Pressure limiter does not work.	a) Broken limiter spring.	a) Replace the spring.
	b) Limiter blocked open.	b) See 2-3 d and 1-4 c.
	c) No pressure in the limiter.	c) Incorrectly mounted pilot head. Vent valve malfunction.
5-4 The limiter gives off too much heat.	a) Temperature in the circuit too high.	a) Incorrect pressure setting.
	b) Oil speed too high.	b) Install a limiter with increased capacity.

4.6 Flow Regulators

FAILURE	CAUSE	SOLUTION
6. Flow regulators		
6-1 The regulator does not work.	a) Regulator under stress. (Installation distortion)	a) Loosen the screws, then tighten them evenly.
	b) Faulty compensation spool.	b) Replace the spool, check liners, possible scratching by the particles.
	c) Faulty metering spool.	c) Replace the slide.
	d) Blocked check valve.	d) Check the valve slide/seat and if necessary replace (broken spring?).
	e) Blocked metering spool.	e) Replace the spool. Check condition of the liner, if necessary change valve. Check the calibrated orifices.
	f) Faulty compensating element. (Sealing parts)	f) Replace.

FAILURE	CAUSE	SOLUTION
	g) Blocked metering valve.	g) Check the cleanliness of the circuit, check calibrated orifices.
	h) Broken spring.	h) Replace the spring.
	i) Corrosion.	i) If necessary clean or change.
	k) Regulator error.	k) Install an appropriate regulator.

4.7 Anti-shock Valves

FAILURE		CAUSE	SOLUTION
7.	Anti-shock valves.		
7-1	The anti-shock system does not work.	a) Broken spring.	a) Replace the spring.
		b) Contaminated circuit.	b) Clean the circuit. Reference to NAS/ISO standard.
		c) Blocked valves or sleeve linings.	c) Replace valve or sleeve liner. Ensure correct re-assembly.

4.8 Actuating Cylinders

FAILURE		CAUSE	SOLUTION
8.	Actuating cylinders		
8-1	Actuating cylinders working too freely.	a) Faulty piston or seals.	a) Service or replace.
		b) Excessive oil leak on the rod guide.	b) Check the drive/head, replace the faulty parts.
		c) Pressure too high or too low.	c) Check the operating pressure at no load and with load to control the friction in the joints.
8-2	Blocked actuator.	a) Other causes.	a) See 2-1 b, f and g.
8-3	Unsteady operation.	a) Barrel of the cylinder not circular.	a) Rework the barrel or replace. Barrel worn or scratched by the interaction of the rod and the piston in operation.
		b) Load variations.	b) Install a counterbalance valve and check value.
		c) Pressure variation (Stick-slip).	c) Check the circuit (12-1). At slow speed examine the tightening of the joints. (Mechanical) (Pulsating feed)

4.9 Filters

FAILURE	CAUSE	SOLUTION
9. Filters		
9-1 Weak filtering.	a) Mesh too coarse.	a) Install a filter with a finer mesh.
	b) Blocked filter, oil by passes.	b) Clean filter and or circuit if necessary. Analyze the reason for the appearance of particles on the filter element-ref NAS/ISO standard
	c) Error in installation.	c) Check the flow direction.
	d) Damaged magnetic elements.	d) Install new magnetic elements.
	e) Clogged elements.	e) Clean, otherwise nstall new elements. Analyze the reason for the presence of particles on the filter element.
	f) External leak from piping (threads?).	f) Seal using Teflon or Loctite.
	g) Mistake in the circuit.	g) Modify the circuit to position the filter in the right place, if possible with a 10 μm mesh.

4.10 Tanks

FAILURE	CAUSE	SOLUTION
10. Tanks		
10-1 Oil pollution.	a) Contaminated access to reservoir.	a) Replace the seals ensuring airtight.
	b) Contaminated circuit.	b) Drain and flush the circuit. Have the particles analyzed to determine their nature/origin. Ref NAS/ISO standard.
	c) Incorrect air filter.	c) Install a better adapted filter.
	d) Faulty air filter.	d) Replace the filter or clean the element.
	e) Clogged circuit and tubes.	e) Clean and scour then clean them again and flush the circuit, watch for any particles on the filter elements.
10-2 Aeration.	a) Oil level too slow.	a) Top-up reservoir after de-foaming.
	b) Non-filled circuit.	b) After putting it into operation fill the circuit (leaks); after de-foaming.

FAILURE	CAUSE	SOLUTION
	c) Poor reservoir design.	c) Install a decanting baffle plate in the tank.
	d) Return pipe flowing above oil level.	d) Install pipe below the oil level.
	e) Cavitation.	e) Too great a depression in suction. Check piping length/dia and the strainer filtering capacity. (See 1-2 f and g)
	f) Oil of poor quality.	f) Follow the advice to operators (for special fluids consult manual). Ref HF Standards.
	g) Too weak a depression by the breather.	g) Modify the tank's air entry system (see 1-1 m.)
	h) Poor assembly of the return pipe.	h) Mounting of a tee on the return pipe-Venturi effect if branch is not tight.
10-3 Temperature too high.	a) No cooling system.	a) Install a cooler or modify the tank for thermal exchange surface.
	b) Poorly adapted cooler.	b) Increase capacity of the cooler/tank thermal exchange surface.
	c) Thermal exchange surface too small.	c) Increase the thermal exchange surface.
	d) Ambient temperature too high.	d) Change the position of the tank or install a cooler.
	e) Circuit too close to the source of heat.	e) Check the distance between the tank and the source of heat and if necessary mount an isolation shield. Make reference to safety.
	f) Pressure in the circuit too high.	f) Modify the initial pressure.
	g) Misinterpretation of the installation.	g) Modify the installation.
	h) Faulty elements in the circuit.	h) Replace these elements.
	i) No oil level indicator, therefore impossible to check.	i) Install a sight level gauge or method of dipstick.

4.11 Couplings

FAILURE	CAUSE	SOLUTION
11. Couplings		
11-1 Coupling becomes hot.	a) Poor axial alignment.	a) Accurately align the coupling/pump/ drive mechanism.
	b) Electrical fault. (Electrical coupling)	b) Repair defect.
	c) Coupling error.	c) Correct coupling?
	d) Rigid coupling.	d) Install a more flexible coupling.
	e) Faulty bearings.	e) Replace the bearings.
	f) Eccentric coupling.	f) Balance the coupling or replace.
	g) Coupling misaligned or loose.	g) Loosen the nuts then tighten them evenly.

4.12 Piping

FAILURE	CAUSE	SOLUTION
12. Piping		
12-1 Vibrations	a) Badly secured piping.	a) Improve anchorage.
	b) Pressure variations in the circuit.	b) Check the link between the pumps and the valves (flex hose strain).
	c) Incorrect alignment.	c) See 11-1 a.
	d) Resonance in the hollow housing. (Chassis plates etc.)	d) Use anti-noise system for the plates, cover the chassis with concrete during sealing.
	e) Anti-shock circuit ?	e) Revise circuit.
	f) Foaming in the circuit.	f) Look for the cause. (See also 10-2 c, d, e, g and h)
	g) Instability of the relief valve.	g) Check the relief valves (See 5-1).
	h) Pulsating flow from the pump (ripple).	h) Check pulsating flow (oscilloscope) and if necessary change the pump.
	i) Air in the circuit.	i) Badly drained circuit. (See also 12-1 f and g)
12-2 Insufficient water-air tightness.	a) Seals incorrectly mounted.	a) Following instructions remount the seals.
	b) No seals.	b) Mount the seals.
	c) Faulty seals.	c) Replace the seals.

FAILURE	CAUSE	SOLUTION
	d) Loose connections.	d) Tighten the connections.
	e) Faulty installation of the pipes.	e) Refer to the mounting instructions.
12-3 Contamination.	a) Circuit not cleaned.	a) Clean the circuit. (See also 1-2 n.)
	b) Circuit not scoured.	b) Scour and clean the circuit again.
	c) Scale deposits on pipes.	c) Descale the pipes, clean the circuit and reassemble the pipes.
	d) Poor welding (scale deposits coming off gradually).	d) Check the welding points and follow the assembling instructions. (See also 10-1 e)

4.13 Accumulators

FAILURE	CAUSE	SOLUTION
13. Accumulators		
13-1 The accumulator is not functioning.	a) Damaged or porous.	a) Return to the supplier for repair. NOTE: Specify the type of fluid so that bladder is compatible.
	b) Pressure in the circuit too low.	b) Increase the initial pressure. Check pre-charge.
	c) Pressure differential to close.	c) Increase differential.
	d) Faulty linings and joints. (Piston accumulators.)	d) Replace the lining or joints.
	e) Error in assembling the accumulator	e) Follow the instructions for assembling. Service or replace.
13-2 Accumulator giving off too much heat.	a) Excessive flow rate.	a) Install a restriction valve and reduce the speed.
	b) Pressure too high.	b) Reduce the initial pressure.
	c) Dirty circuit.	c) Clean or scour the circuit then clean it again. (See also 1-2 n)

4.14 Oil Cooler

FAILURE	CAUSE	SOLUTION
14. Oil Cooler.		
14-1 Poor cooling.	a) Starting temperature of the cooler too high.	a) Revise oil/water flow path and rate.
	b) Clogged water circuit.	b) Clean the water circuit.

FAILURE	CAUSE	SOLUTION
	c) Fan power too weak.	c) If possible increase the power or consult the supplier.
	d) Inadequate cooler.	d) Change the cooler.
	e) Defect in the cooler water supply.	e) Check the water inlet. (Choose the piping closest to the pumping point.)
	f) Faulty fan.	f) Repair the fan.
	g) Error in the manufacturing of the cooler.	g) Change the cooler.
	h) Uprated system.	h) Check that the cooler is adequate for uprated system.
	i) Heat exchange capacity cooler insufficient and recycling of the oil too rapid.	i) Use a larger cooler or increase the capacity of the tank.
14-2 Water emulsion in the oil.	a) Fractured cooler.	a) Repair the circuit and completely drain out the oil several times if necessary until there is no trace of water.
	b) Condensation phenomenon. Cooler has its core incorporated into the tank.	b) To avoid this phenomenon check the circuit; the condensation appears when the water temperature is very low and when there is little oil in the tank. (Core exposed)

4.15 Miscellaneous

FAILURE	CAUSE	SOLUTION
15. Miscellaneous		
15-1 Contamination.	a) Poor quality pipework scale.	a) Check assembly instructions. Piping without annealing, scouring welding and greasing.
	b) Exposed oil passages.	b) Place the covers during assembling, flush before start-up. (See also 1-2 n)
	c) Faulty filtering.	c) Improve the filtering. (see 9-1)
	d) Cylinder rods allow dust to enter.	d) Install scraper joints, anti-dust collars or boots.
15-2 Aeration.	a) Air in the circuit.	a) Evacuate air from the circuit (see also 10-2).
	b) Cavitation.	b) See 10-2 e.
	c) Return line above the oil level.	c) Change to below the oil level, adapt oil diffusers.

FAILURE	CAUSE	SOLUTION
15-3 Variation in temperature.	a) No thermostat on the cooler.	a) Install a thermostat.
	b) Erratic water flow of the cooler.	b) Check the water circuit, thermostat, electrical circuit, water temperature.
15-4	a) Suspect pressure control system.	a) Revise system.
	b) Faulty thermostat.	b) Repair/replace.
	c) Current leak. (Humid atmosphere)	c) Check resistance, use component protection cover.
	d) Faulty electrical circuit.	d) Check the circuit.

4.16 Some Typical Component Failures, Pumps and Motors

The preceding list of faults is supported by many photographs of typical failures and in this section an attempt is made to convey just a few of these examples and specifically for pumps and motors. This section also focuses on just a few aspects to avoid reproduction of many excellent brochure photographs. The importance of mechanical failures and their causes should therefore not be overlooked and the reader should refer to the greater volume of information associated with this extract. Figures 4.1 and 4.2, for example, shows just some of the possible consequences of mechanical failures.

System pressures are constantly rising resulting in increased pressure overshoot. The effects on pumps, whichever technology is used, are bad. Consider then two different categories, "Instant pressure overshoot" and "Cycled overpressurisation". The final consequences of these are the same, that is, the failure of components. It will be seen that the failing components are damaged differently if it is instant pressure overshoot or cycled overpressurisation. The valves and the pipes rigidity and length around the pumps have a great impact on these pressure peaks; it can be a system problem or a valve problem that opens the main security valve. The pump is, or is not, protected by a check valve. The fact is that the pressure rises over the initial settings or designed settings. This problem is mainly seen when the valves tend to open (or close for a check valve) too slowly. These pressure peaks can reach 2 to 5 times the pressure settings and with a transient response that may not be detectable with standard dial gauges. When the check-valve closes itself too slowly, the flow will come backwards into the pump. This problem will be seen in the cycled overpressurisation. Figures 4.3 and 4.4 show some results of overpressurisation.

Instant pressure overshoot is brutal in the sense that the mechanical strength of the material is exceeded with devastating results, as seen in Figure 4.3.

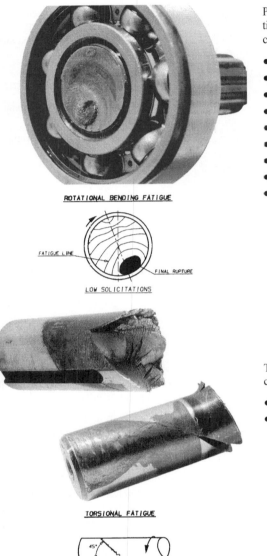

Perpendicular, over-centre rotational bending fatigue rupture, with causes such as:

- Bad alignment
- Out of squareness
- Unbalanced coupling
- Too high radial loads
- Non-homokinetic
- Too large an inertia
- Bracket chassis deformation
- Hose strain force
- Bad shaft/coupling link

Twisted torsional rupture, with causes such as

- Fretting corrosion
- Over torque limits

Figure 4.1. Some typical shaft failures for a vane pump

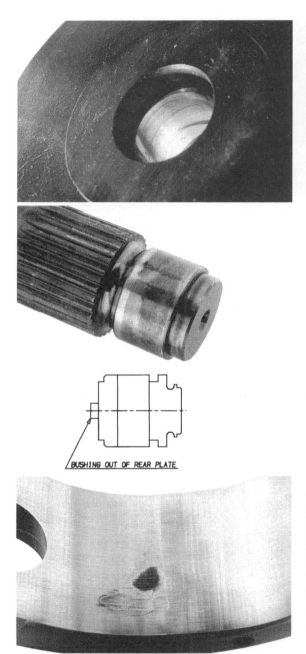

Bushing/bearing problems

Front or back bearing/bushing with heavy wear, and with causes such as:

- Shaft problems
- Bad shaft/coupling connection

Bushing welded on the shaft, and with causes such as:

- Shaft problems
- Bad shaft/coupling connection

Back bushing moving out of the rear port plate, and with causes such as:

- Shaft problems
- Bad shaft/coupling connection

Cam ring marked by the rotor on the smallest of the diameter. If the contact between the rotor and the cam ring is important, it will transform the hardness of the cam ring and create local tensions (cracks). Possible causes are:

- Shaft problems
- Bad shaft/coupling connection

Figure 4.2. Some consequences of mechanical failures for a vane pump

a) cracks or rupture of the pressure plate   b) cam ring cracked

c) cracks or rupture of the rotor   d) shaft broken, perpendicular clean cut

Figure 4.3. Some consequences of instant pressure overshoot for a vane pump

In the presence of aeration the vanes in a vane pump become unbalanced due to the drastic reduction in the fluid effective bulk modulus. The vane, usually hydrostatically balanced, will move sideways with such erratic movements that the vanes will destroy their lubrication film of oil that links them to the side plates. The hardened pieces of the vanes will start to wear the side plate in die cast or ductile iron. The marks will start in the discharge area and, depending on the quantity of air, a groove will probably form. *The most noticeable fact will be an unusual noise level*. Figure 4.5 shows some common failures.

Cavitation/de-aeration occurs when a depression arises at the suction port. The gas (combustible) and aromatic essences dissolved in the fluid (6 to 7%) will evaporate. Depending on the type of fluid, this de-aeration will occur at pressures between 100 and 150 mm of HG (around −0.2 bar). Under this depression (or vacuum), small bubbles of diameter 0.2 to 0.3 mm will be formed. The natural appearance of oil is translucent. Under cavitation and because of these small "bubbles", the fluid will have a "cloudy" appearance. Depending on the value of the vacuum, the quantity of suspended bubbles will be more or less important. As these bubbles have a small diameter, they will reach the surface of the oil tank

Rupture/cracks of cam ring

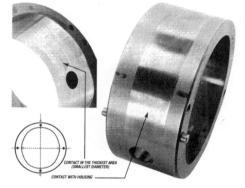

Rotor/cam ring contact in the smaller diameter

Shaft internal splines worn

Shaft rupture. Torsional fatigue ruptures perpendicular, few cycles but very high torque

Shaft twisted often under high cycling

Figure 4.4. Some consequences of cycled pressure overshoot for a vane pump

Under very severe aeration the port plates can be deeply marked from the suction area to the outlet area. The width of the groove ist then the width of the vane.

In very heavy aerated conditions the vane is so unbalanced that ist can sometimes break.

Noise will be evident

Figure 4.5. Some consequences of aeration within a vane pump

very slowly (bad de-aeration characteristics). As an example, 100 litres of an oil foamed by cavitation will take 4 hours to become translucent again. When the fluid reaches local hot temperatures and is compressed (at the "critical pressure"), these bubbles implode and create a shock wave, known as the diesel effect, the impact of these "combustion explosions", will create erosion in the shape of craters (cavities) when located near a metallic surface. These detached metallic particles are very likely to cause, on a medium term base, a seizure between the pump's moving parts.

This problem can be caused by a number of bad features such as:

(i)   suction strainer that can become clogged by a foreign contaminant or clogged by a high-viscosity fluid

(ii)   large pressure drop, an overlong inlet hose, an undersize diameter inlet, inlet tube in the tank too close to the panel of the tank

(iii) inlet tube in the tank with undersize suction surface creating local turbulence (de-aerating the fluid). Cut the tube at an angle to increase this suction surface and avoid local high velocities

(iv)   too high or too low inlet velocity (0.5 to 1.9 m/s is the velocity required)

(v)   tank too far away from the pump (horizontally or vertically)

(vi)   excessive shaft speed

(vii) air filter on the tank clogged or not well dimensioned, generating a vacuum in the tank

(viii) reservoir oil level too low compared to the suction level (when all cylinders are extended, for example)

(ix)   inlet tube in front of the return line (amplifying the foaming)

(x)   tank too small (high velocity in the tank)

(xi)   bad de-aeration capabilities of the oil and of the tank. Baffles can help by "pushing" the air to the surface. If the "vein flow" is too rapid and if no baffle is there to bring these bubbles to the surface, they will go back to the inlet area. This air in the pump will deteriorate it.

(xii) bad filtration dimension on return line. Under-dimension will increase the velocity and de-aerate the oil.

Some effects due to cavitation are shown in Figure 4.6.

Consider now some consequences of particle contamination within a vane pump. "Unlike a lot of different technologies, Denison Hydraulics vane units do not generate pollution". Contamination is an important topic and a lot of progress has been made concerning the cleanliness of the fluid. However, pollution by particles remains one of the greatest causes of pump destruction. The consequences are either rapid wear or premature breakdown (large size particles over 25 μm). In a hydraulic circuit, the pump is usually the most sensitive unit to pollution and therefore will be the first component to fail. The main particles are made up of metallic oxide, silica, carbon and organic materials. Particles can originate from a number of sources.

- A common large particle is metallic oxide from welding burrs when welded piping has not been cleaned properly.
- Silica comes from the surrounding dust. This dust will enter into the system through the seals of cylinders, through air intakes (absence of air filters), from a dirty environment if the tank is not properly sealed.

Some consequences of particle pollution are shown in Figure 4.7.

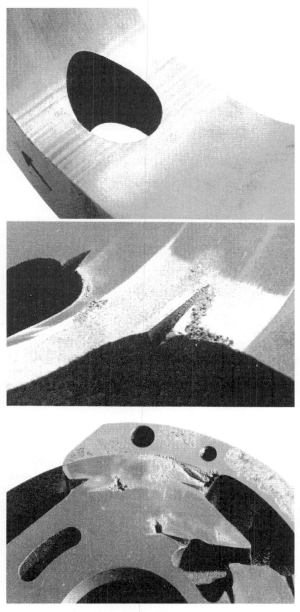

Ripples on the cam ring: the vanes are hydrostatically balanced to avoid excessive loads on the vane lips. During the suction cycle, the pin compensates the out of balance load due to the cam profile. When the depression is over the design limits, the vane bounces, creating ripples on the cam ring profile. The depth of these marks is proportional to the strength of the depression.

Craters: erosion craters are sometimes difficult to observe as the pump may have already seized. Craters come from erosion, caused either by an explosion/implosion, due to depressurisation. When the fluid trapped between the two vanes is sucked in with a certain percentage of air in suspension, an explosion can occur. When this trapped volume is compressed, these air bubbles explode and create craters in the side plates in the area between the suction port and the pressure port, around the pressure bleed slots.

Figure 4.6. Some consequences of cavitation within a vane pump

The vane surface.
The film of oil between the vanes and the rotor being contaminated, there will be a rubbing effect in this area. These rubbings marks (pollution marks) will be vertical and of the height of the vanes' translation (displacement).

Cam ring.
Between the vane lip and the cam ring, the film of oil is contaminated. This will wear the inner surface of the cam ring.

Rotor/side plates
When the particles in suspension in the fluid are greater than half the clearance between the rotor and cam ring, seizure occurs in the peripheral diameter of the rotor and the port plate

Figure 4.7. Some effects of particle pollution

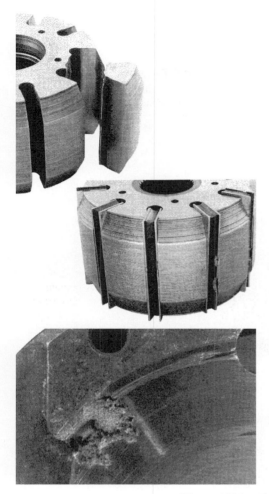

Rotor

The rubbing effect will also appear in-between the side of the rotor and the side plates. This will create a torque between the two vanes laps. This causes a reasonably high level of fatigue in the materials' weakest area, between the two bulb slots of the rotor. If this fatigue level exceeds the design limits, this portion of the rotor will break.

Rotor/side plates/vanes

Big contamination particles damage and usually seen on the port plates (blocked in the slots) or/and on the top of the vanes/rotor. Each time, they will have an effect on the vane lips, either on the top either on the sides. The "rubbing" action will either destroy the vane lips or weld the vane to the rotor, break the cam ring.

Side plates

Another sign of contaminated oil is some possible erosion craters on the port plates at the inlet/suction bleed slots area. These erosion craters would come from the abrasive fine particles in a local high velocity area.

Figure 4.7. (continued)

Depending on the size of the particles, the consequences range from a gentle ground finish on the vane lips, cam surface and side plates, to total destruction of the cartridge.

When considering *axial piston machines* the effect of aeration and particle contamination gives end results similar to those previously described for vane machines. For example, Figures 4.8 → 4.11 show further practical evidence.

Metal or mineral particles are induced into the spherical chamber during the combined action of the negative force during depression (suction) and then become trapped when the spherical chamber is blocked (angular displacement) when the pressure is applied. The piston sphere presses the particles into the bronze shoe creating accelerated wear. Analysis and experience have shown that these foreign elements turn the shoe spherical zone into a diamond grinder which causes very rapid wearing of the piston spherical head, which is made of specially selected steel, see Figure 4.12.

Figure 4.8. Seizing the barrel wearplate face on the port plate with signs of heating and attributed to poor suction

Figure 4.9. Excessive vacuum creating shoe detachment from the swash plate resulting in a hammering effect causing a breakdown of the shoe periphery

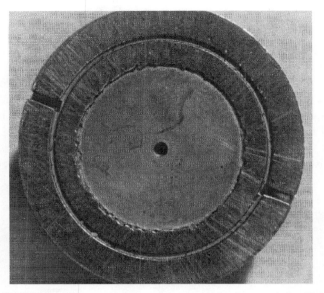

Figure 4.10. Radial grooves caused by particle contamination carried with the inherent leakage flow across the shoe land

Figure 4.11. Destruction of the sealing ring around the chambers affects the hydrodynamic imbalance of the shoes and results in rapid wear of the shoes as well as the swash plate

In addition to the wear effect shown in Figure 4.12, wear of the piston occurs and, due to the orientation of the piston within the barrel bore, increased wear is evident at the shoe-end of the piston, as shown in Figure 4.12 and re-viewed in Figure 4.13.

Figure 4.14 illustrates axial piston pump swash plate damage due to slipper lift. This can be reflected as a rolling effect from the slippers, and the condition shown in Figure 4.14(a) is retrievable in the sense that the swash plate can be reground. This is not the case shown in Figure 4.14(b) where repetitive slipper bounce has resulted in severe damage to the swash plate. The slipper could have separated from the piston or bounce could have resulted from unstable feedback control loop dynamics.

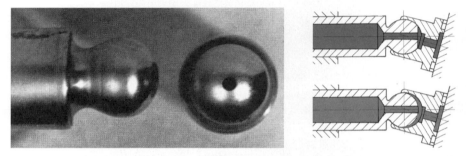

Figure 4.12. Particle contamination creating wear at the shoe spherical zone and also affecting the balance force due to increased effective pressure area

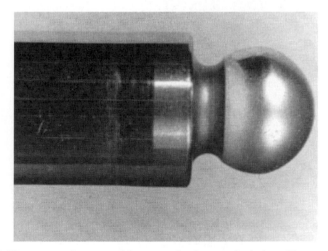

Figure 4.13. Piston wear due to particle contamination, particularly evident at the shoe-end of the piston

a)   Modest wear, swash plate retrievable

b)   Severe damage, irretrievable

Figure 4.14. Axial piston pump swash plate damage due to slipper bounce

Further Reading

Often a fluid power topic does not fit into an obvious category from the broad chronological list that follows. The reader may have to search through one or more of these categories to locate a specific topic.

General Condition Monitoring Papers and Text Books

1   Witt K and Schlosser WMJ. Thermodynamic measurements on hydraulic components. ICMES Conference, Paris, 1977
2   Watton J. Monitoring the wear characteristics of a positive displacement vane pump. IMechE Pumps for difficult liquids, 1985, 43–47
3   Mbari PN and McCandlish D. Reliability and Fault Tree Analysis in hydraulic systems. 7th International Fluid Power Symposium, Bath, UK, 1986, 303–311
4   Hunt TM. A review of Condition Monitoring techniques applicable to fluid power systems. 7th International Fluid Power Symposium, Bath, UK, 1986, 285–294
5   South CJ. Aircraft hydraulic supply circuit health monitoring. 7th International Fluid Power Symposium, Bath, UK, 1986, 295–302
6   Ding G and Dayue H. Monitoring and diagnosis of typical faults in hydraulic systems. Condition Monitoring '87, edited by MH Jones, Pineridge Press Ltd, 1987, 593–603
7   Sargent CM, Burton RT, Westman RV. Expert systems and fluid power. Fluid Power 8, published by Elsevier Applied Science, 1988, 423–441
8   Pippenger JJ, Zero Downtime Hydraulics. Amalgam Publishing Company, courtesy of Sun Hydraulics Corporation USA, 1989
9   Yang H. Temperature analysis in hydraulic systems. Proc 9th International Symposium on Fluid Power, published by STI Oxford, 1990, 207–221
10   Martin KF and Thorpe P. Coolant system health monitoring and fault diagnosis via health parameters and fault dictionary. The International Journal of Advanced Manufacturing Technology, 1990, 66–85
11   Watton J, Condition Monitoring and Fault Diagnosis of Fluid Power Systems. Published by Ellis Horwood, 1992
12   Atkinson R M et al. Automated fault diagnosis for hydraulic systems, Part 1: fundamentals. Proc IMechE Part I Journal of Systems and Control Engineering, 1992, 206, 207–214
13   Hogan P A et al. Automated fault diagnosis for hydraulic systems, Part 2: applications. Proc IMechE Part I Journal of Systems and Control Engineering, 1992, 206, 215–224

14   Darling R and Tilley D G. Progress towards a general purpose technique for the condition monitoring of fluid power systems, Proc IMechE Conference on Aerospace Hydraulics and Systems, London, 1993, 47–55

15   Ramden T, Weddfelt K, Palmberg J-O. Condition Monitoring of Fluid Power Pumps by Vibration Measurement, 10th International Conference on Fluid Power-The future for Hydraulics, Brugge, Belgium, MEP Publications Ltd, 1993, 263–276

16   Wen Z et al. Research on condition monitoring failure in axial piston pumps. Proc 2st JHPS International Symposium on Fluid Power, Tokyo, 1993, 229–234

17   Watton J, Lucca-Negro O, Stewart JC. An on-line approach to fault diagnosis of fluid power cylinder drives systems. Proc IMechE Journal Systems and Control Engineering, 1994, Vol 208, 249–262

18   Pouliezos A D and Stavrakakis G S, Real Time Fault Monitoring of Industrial Processes,. Kluwer Academic Publishers, 199

19   Hehn A. Fluid Power Design Handbook. Marcel Dekker, 1994

20   Hunt T M, Condition Monitoring of Mechanical and Hydraulic Plant. Published by Chapman and Hall 1996

21   Rao B K N, Handbook of Condition Monitoring. Published by Elsevier Advanced Technology, 1st edition 1996

22   Condition Monitor, An International Newsletter. Published monthly by Coxmoor Publishing Company UK.

23   Watton J and Stewart JC. Co-operating expert knowledge and artificial neural networks for fault diagnosis of electrohydraulic cylinder position control systems. Proc 3rd JHPS International Symposium on Fluid Power, Yokohama, 1996, 217–222

24   Kunimoto E and Ogawara T. Cavitation detection in the oil-hydraulic equipments. Proc 3rd JHPS International Symposium on Fluid Power, Yokohama, 1996, 461–466

25   Ramden T, Krus P, Palmberg J-O. Reliability and sensitivity analysis of a condition monitoring technique. Proc 3rd JHPS International Symposium on Fluid Power, Yokohama, 1996, 567–572

26   Hogan PA et al. Automated fault tree analysis for hydraulic systems. Trans ASME, Journal of Dynamics Systems, Measurement and Control, Vol 118, 1996, 278–282

27   Bull D R, Stecki J S, Edge K A, Burrows C R. Failure Modes and Effects Analysis of a valve controlled hydrostatic drive. Proc 10th Bath International Fluid Power Workshop, Research Studies Press Ltd 1997, 131–143

28   Andrews J and Henry J. A computerised fault tree construction methodology. Proc IMechE, Part E, Journal of Process Mechanical Engineering, 1997, 211, 171–183

29   Le T T, Watton J and Pham D T. An artificial neural network based approach to fault diagnosis and classification of fluid power systems, Proc IMechE, Part I, Journal of Systems and Control Engineering, 1997, Vol 211, 307–317

30   Rinkinen J, Laukka J, Ahola E. Condition diagnosis of servovalve in oil hydraulic servo system of hot strip mill. Proc 5th Scandinavian Conference on Fluid Power, Linkoping, Sweden, 1997, 429–444

31   Mitcheel R J and Pippinger J J, Fluid Power Maintenance Basics and Troubleshooting. Marcel Dekker, 1997

32   Stoneham D, Maintenance Management and Technology Handbook. Elsevier Advanced Technology, 1st edition, 1998

33   Planning and Design of Hydraulic Power Systems. Mannesmann Rexroth Hydraulic Trainer Vol 3, Mannesmann Rexroth, Postfach 340, D8770 Lohr am Main, Germany

34   King R and Hurst R (ed.). King's Safety in the Process Industries. Published by Arnold, Second Edition 1998

35   Protecting the Environment in Process Industry. 1st International Conference, Professional Engineering Publications, 1998

36   Crowther WJ et al. Fault diagnosis of a hydraulic actuator circuit using neural networks-an output vector space classification approach. Proc IMechE, Part I, Journal of Systems and Control Engineering, 1998, Vol 212, 57–68

37   Reeves CW, Vibration.. Published by Coxmoor, 1998

38   Le T T, Watton J and Pham D T. Fault classification of fluid power systems using a dynamics feature extraction technique and neural networks. Proc IMechE, Part I, Journal of Systems and Control Engineering, 1998, Vol 212, 87–97

39   Khoda T and Inque K. Knowledge acquisition for failure diagnosis of hydraulic systems using bond graph model, Power Transmission and Motion Control, PTMC99, Bath University, UK. Professional Engineering Publications, 1999, 123–134

40   Freebody N and Watton J. A time encoded signal processing approach to fault classification of an electrohydraulic pressure control system and its application to a hot steel strip rolling mill. Proc IMechE, Part I, Journal of Systems and Control Engineering, 1999, Vol 213, 407–426

41   Martin KF and Marzi MH. Diagnostics of a coolant system via neural networks. Proc IMechE Part I Journal of Systems and Control Engineering, Vol 213, 1999, 229–241

42   Rinkinen, J., Laukka J. Karinen K., Online Condition Diagnosis of Hydraulic Filter in Hot Strip Mill. The Sixth Scandinavian international conference on Fluid Power, SICFP'99, Tampere, Finland,1999, 635–649.

43   Freebody N and Watton J. CARCODE for time encoded signal processing and fault trending of the work roll bending hydraulic control system on a 7-stand hot steel strip rolling mill. Proc 4th JHPS International Symposium on Fluid Power, Tokyo, 1999, 487–492

44   Noguchi E and Nagata K. Detection of vibration path in hydraulic power steering system. Proc 4th JHPS International Symposium on Fluid Power, Tokyo, 1999, 507–512

45   Gale KW and Watton J. A real-time expert system for the condition monitoring of hydraulic control systems in a hot steel strip finishing mill. Proc IMechE, Part I, Journal of Systems and Control Engineering, 1999, Vol 213, 359–374

46   Kohda T and Inque K. Knowledge acquisition for failure diagnosis of hydraulic systems using system bond graph model. Power Transmission and Motion Control 1999, 123–134. Professional Engineering Publications Ltd

47   Watton J and Freebody N. A time encoded signal processing/neural network approach to fault classification of an electrohydraulic control system. International Journal of Fluid Power, 2000, No 2, 59–66

48   Rinkinen, J., Elo, L. & Laukka, J. 2000. Application of Diagnostics in the Condition Monitoring of the Hydraulic Gauge Control System, Case: Hot Strip Rolling Line, Rautaruukki Steel. In: Lahdelma, S. & Hietala, M.(eds.). Proceedings of Maintenance, Condition Monitoring and Diagnostics - International Seminar, Oulu, Finland. Oulu. Pohto, Merityrsky Oy, 2000, 79–89.

49   Mourre D and Burton R. Investigation of a neural network/statistical condition monitoring technique for a proportional solenoid valve. Power Transmission and Motion Control 2001, 119–134. Professional Engineering Publications Ltd

50   Da Silva C. Knowledge management and fluid power. Proc 2nd International Workshop on computer software for design, analysis and control of fluid power systems. VSB Technical University of Ostrava/Fluid Power Net International, Czech Republic, 2001, 167–174

51   Bailey S and Watton J. A neural network approach to transmission line modelling and fault diagnosis of a hydraulic press control system. Proc IMechE, Part I, Journal of Systems and Control Engineering, 2002, Vol 216, 357–367

52   Stecki JS, Conrad F, Beng OH. Software tools for automated Failure Modes and Effects Analysis (FMEA) of hydraulic systems. Proc 5th JFPS International Symposium on Fluid Power, Nara, Japan, 2002, 889–894

53   Meindorf T and Seyfert C. First results of online condition monitoring of hydraulic fluids. Power Transmission and Motion Control 2002, 115–126. Professional Engineering Publications Ltd

54   Kim H-E et al. Fully automated diagnostic system development for hydraulic motor performance. Proc 5th JFPS International Symposium on Fluid Power, Nara, Japan, 2002, 549–554

55   Rusanen, H. & Rinkinen, J. 2004. Model Oriented Condition Monitoring of Hydrostatic Transmission. Codina Macia, E. et al. (Eds.). Proc 3rd FPNI-PhD Symposium on Fluid Power, Terrassa Spain, 2004, 489–500

56   Scott MJ, Load Monitoring. Published by Coxmoor, 2004

57   Scheffer C and Girdhar P, Practical Vibration Analysis & Predictive Maintenance. Published by Elsevier, 2004

58   Macdonald D.,Practical Machinery Safety. Elsevier, 2004

59   Macdonald D.,Practical Industrial Safety, Risk Assessment and Shutdown Systems. Elsevier 2004

60   Laitinen L et al. Ageing dependant characteristic changes of a pressure control valve for model-based condition monitoring in fluid power. Power Transmission and Motion Control Workshop, PTMC 2004, Bath, 285–295. Published by Professional Engineering Publications Ltd

61   Jouppila V, Kuusisto J, Ellman A. A model based method for condition monitoring of a proportional valve. Power Transmission and Motion Control Workshop, PTMC 2004, Bath, 309–317. Published by Professional Engineering Publications Ltd

62   An L and Sepehri N. Hydraulic actuator leakage fault detection using extended Kalman Filter. International Journal of Fluid Power 6, 2005, No 1, 41–51

Other Fluid Power Text Books and Manuals

63   Thoma JU. Hydrostatic Power Transmission. Trade and Technical Press Ltd 1964

64   Merritt HE. Hydraulic Control Systems. John Wiley and Sons 1967

65   Korn J. Hydrostatic Transmission Systems. Intertext Books 1969

66   Guillon M. Hydraulic Servo Systems. Butterworths 1969

67   Goodwin AB. Fluid Power Systems—Theory, Worked Examples and Problems. The Macmillan Press 1976

68   Pippenger J and Hicks T. Industrial Hydraulics. McGraw-Hill Kogakusha 1979

69   McCloy D and Martin HR. Control of Fluid Power—Analysis and Design. Ellis Horwood Ltd 1980

70   Lambeck RP. Hydraulic Pumps and Motors—Selection and Application for Hydraulic Power Control Systems. Marcel Dekker 1983

71   Reed EW and Larman IS. Fluid Power with Microprocessor Control. Prentice Hall International 1985

72   Banks DS and Banks DD. Industrial Hydraulic Systems. Prentice Hall 1988

73   Pinches MJ and Ashby JG. Power Hydraulics. Prentice Hall 1988

74   Anderson WR. Controlling Electrohydraulic Systems. Marcel Dekker 1988

75   Rohner P i) Industrial Hydraulic Control, ii)Industrial Hydraulic Control—workbook, Teacher's Edition, iii)Industrial Hydraulic Control—workbook. AE Press Melbourne 1988

76   Watton J. Fluid Power Systems—Modelling, Simulation, Analogue and Microcomputer Control. Prentice Hall 1989

77   Kleman A. Interfacing Microprocessors in Hydraulic Systems. Marcel Dekker 1989

78   Thoma JU. Simulation by Bomdgraphs. Springer-Verlag 1990

79   Mannesmann Rexroth series on fluid power. Second issue 1991,Vol 1 Basic Principles and Components of Fluid Power Technology, Vol 2 Proportional and Servo Valve Technology, Vol 3 Planning and Design of Hydraulic Systems, Vol 4 logic Element Technology, Vol 6 Hydrostatic Drives with Control of the Secondary Circuit

80   Vickers Industrial Hydraulics Manual 1992Vickers Closed loop Electrohydraulics Systems Manual 1992

81   Watton J. Condition Monitoring and Fault Diagnosis in Fluid Power Systems. Ellis Horwood 1992

82   Norvelle FD. Fluid Power Technology. West Publishing Company 1995

83   Martin H. The Design of Hydraulic Components and Systems. Ellis Horwood 1995

84   Yeaple F. Fluid Power Design Handbook. Marcel Dekker 1995

85   Turner IC. Engineering Applications of Pneumatics and Hydraulics. Arnold 1996

86   Hodges P. Hydraulic Fluids. Arnold 1996

87   Water Hydraulics Control Technology. E Trostmann. Published by Marcel Dekker Inc, 1996

88   Konami S and Nishiumi T. Hydraulic Control Systems (in Japanese). TDU 1999

89   Ivantysyn J and Ivantysynova M. Hydrostatic Pumps and Motors. Akademia Books International, New Delhi 2001

90   Paszota Z. Aspects Energetiques Des Transmissions Hydrostatiques. Wydawnictwo Politechniki Gdanskiej 2002

91   Stecki JS and Garbacik AJ. Design and Steady State Analysis of Hydraulic Control Systems. Fluid Power Net Publications 2002

92   Nakayama Y and Boucher RF. Introduction to Fluid Mechanics. Butterworth Heinemann 2002

93   Chapple PJ. Principles of Hydraulic Systems Design. Coxmoor Publishing Company 2003

94   Younkin G W. Industrial servo control systems, fundamentals and applications. Marcel Dekker 2002

95   The safe use and handling of flammable liquids. Health and Safety Executive, UK 2003

96   Parker Hannifin Design Engineers Handbook, Volume 1 Hydraulics 2001

97   Nervegna N. Oleodinamica E Pneumatica, Politecnico di Torino Fluid Power Research Laboratory Course Notes. Vol 1 Sistemi, Vol 2 Componenti, Vol 3 Esercitazioni

98   Servo and Proportional Systems Catalog. Moog Corporation Inc, East Aurora, New York 14052–0018

Valves and Associated Sub-Components

99   Lee SY and Blackburn JF. Contributions to Hydraulic Control-steady state axial forces on control valve pistons. Trans ASME, 1952, 1005–1011

100   Kreith F and Eisenstadt R. Pressure drop and flow characteristics of short capillary tubes at low Reynolds numbers. Trans ASME,1957, 1070–1078

101   Clark RN. Compensation of steady state flow forces in spool-type hydraulic valves. Trans ASME, 1957, 1784–1788

102   Shearer JL. Resistance characteristics of control valve orifices. IMechE Proc Symposium on recent mechanical engineering developments in automatic control, London 1960, 35–41

103   Stone JA. Discharge coefficients and steady-state flow forces for hydraulic poppet valves. Transactions of the ASME 1960, 144–154

104   Hagiwara T. Studies on the Characteristics of Radial Flow Nozzles (1st report). Bull. Japan. Soc. Mech. Engrs. 1962, Vol 5 No 20, 656–663

105   Takenaka T, Yamane R, Iwamizu T. Thrust of the Disk Valves. Bull. Japan. Soc. Mech. Engrs. 1964, Vol 7 No. 27, 558–566

106   Nikiforuk PN, Ukrainetz PR, Tsai SC. Detailed analysis of a two-stage four-way electrohydraulic flow control valve. Proc ImechE, Journal of Mechanical Engineering Science, Vol 10, 1968, 133–140

107   Mills RD. Numerical solutions of viscous flow through a pipe orifice at low Reynolds numbers. Proc ImechE, Journal of Mechanical Engineering Science, Vol 11, 1969, 168–174

108   Green WL. The poppet valve-flow force compensation. 1970 Fluid Power International Conference, paper 2, S1-S6

109   Urata E. Thrust of poppet valve. Bulletin JSME. 1969, vol 12 No 53, 1099–1109

110   Funk JE, Wood DJ, Chao SP. The transient response of orifices and very short lines. Trans ASME Journal of Basic Engineering, 1972, 483–491

111   de Pennington A, 't Mannetji JJ, Bell R. The modelling of electrohydraulic control valves and its influence on the design of electrohydraulic drives. IMechE Journal of Mechanical Engineering Science, 16, 1974, 196–204

112   Martin DJ and Burrows CR. The dynamic characteristics of an electrohydraulic servovalve. ASME Journal of Dynamic Systems, Measurement, and Control, 1976, 395–406

113   Scheffel G. Dynamisches Verhalten eines directgesteurten Kegelsitzventils unter dem Einfluss der Geometrie des Schliesselementes. Ölhydraulic und pneumatik N 5 1978, 280–282

114   Scheffel G. Dynamisches Verhalten eines directgesteurten Kegelsitzventils unter dem Einfluss der Geometrie des Schliesselementes. Ölhydraulic und pneumatik N 8 1978, 445–448

115   Scheffel G. Einfluss des hydraulischen Schwingungsdämpfers auf das dynamische Verhalten eines Druckbegrenzungsventils. Ölhydraulic und pneumatik N 10 1978, 583–586

116   Nigro FEB, Strong AB, Alpay SA. A numerical study of the laminar viscous incompressible floe through a pipe orifice. Trans ASME, Journal of Fluids Engineering, Vol 100, 1978, 467–472

117   Nakada T and Ikebe Y. Measurement of the axial flow force on a spool valve. The Journal of Fluid Control, Vol 13, 3, 1981, 29–40

118   Takahashi K and Tsukiji T. The unsteady jet from the metering orifice of a spool valve. Bulletin JSME. 1982, Vol 25 No 202, 576–582

119   Watton J. The effect of drain orifice damping on the performance characteristics of a servovalve flapper/nozzle stage. ASME Journal of Dynamic Systems, Measurement and Control. March 1987, Vol 109, 19–23

120   Farr GPR. Brake pressure apportioning valves. Proc IMechE, Vol 201, D3, 1987,193–199

121   Watton J. The design of a single-stage relief valve with directional damping. Journal of Fluid Control, Vol 18, No 2 March 1988, 22–35

122   Kollek W. Kudzma Z. Untersuchung des Einflusses von Konstruktionsparametern auf Strömungserscheinungen in Sitzventilen mit kegelförmigen sperrsystem. Konstruktion 40 1988, 267–271

123   Watton J. The effect of servovalve underlap on the accuracy and dynamic response of single-rod actuator position control systems. Journal of Fluid Control, 18, 1988, 7–24

124   Watton J. The design of a single-stage relief valve with directional damping. Journal of Fluid Control. Vol 18, No 2, March 1988, 22–35

125   Tesar V and Watton J. Hydrodynamics of an idealised poppet valve. Proceedings of FLUCOME '88 Conference. Sheffield, pp 36–39

126   Watton J. Transient analysis of a back pressure-controlled relief valve for controlling the supply pressure rate of rise. Proceedings of 1st JHPS International Symposium on Fluid Power 1989, 373–378

127   Weixiang S. Songnian L, Sihua G. A new technique for steady state flow force compensation in spool valves. Proc ImechE, Part E, Journal of Process Mechanical Engineering, Vol 204, 1990, 7–14

128   Watton J. On linearised coefficients for an underlapped servovalve coupled to a single-rod cylinder. Journal of Dynamic Systems, Measurement and Control, Dec 1990, Vol 112, 794–796

129   Watton J. The stability and response of a two-stage pressure-rate-controllable relief valve. Journal of Fluid Control, Vol 20, No 3, 1990

130   Johnston DN and Edge KA. The impedance characteristics of fluid power components: restrictor and flow control valves. Proc IMechE Vol 205 Part I: Journal Systems and Control Engineering 1991, 3–10

131   Edge KA and Johnston DN. The impedance characteristics of fluid power components: relief valves and accumulators. Proc IMechE Vol 205 Part I: Journal Systems and Control Engineering 1991, 11–22

132   Johnston DN, Edge KA, Vaughan ND. Experimental investigation of flow and force characteristics of hydraulic poppet and disk valves. Proc IMechE Vol 205 1991, 161–171.

133   Petherick PM, Birk AM. State of the art review of pressure relief valve design, testing and modelling. Journal of pressure Vessel Technology. ASME. Vol 113, February 1991, 46–54

134   Washio S. Nakamura Y, Yu Y. Static characteristics of a piston type relief valve. Proc IMechE Vol 213 Part C: Journal of Mechanical Engineering Science, 1991, 231–239

135   Nakano K. Experimental study for the compensation of axial flow force in a spool valve. The Journal of Fluid Control, Vol 21, 1992, 7–26

136   Sallet DW, Nastoll W, Knight RW, Palmer M E, Singh A. An experimental investigation of the internal pressure and flow fields in a safety valve. Winter Annual Meeting, Nov 15–20 1981. Washington DC. Published by ASME 81-WA/NE-19, 1–8

137   Vaughan ND, Johnston DN, Edge KA. Numerical simulation of fluid flow in poppet valves. Proc IMechE Vol 206 1992, 119–127

138   Baylet V. O'Doherty T, Watton J. Reversed flow characteristic of a cone seated poppet valve. Proceedings of the 5th International Symposium on Refined Flow Modelling and Turbulence_Measurements, Paris, Sept 1993, pp 867 -874, published by Presses Ponts et Chaussees.

139   Tsukiji T, Soshino M, Yonezawa Y. Numerical and experimental flow vizualisation of unsteady flow in a spool valve. Proc 2st JHPS International Symposium on Fluid Power, Tokyo, 1993, 379–384

140   Ueno H, Okajima A, Muromiya Y. Visualisation of cavitating flow and numerical simulation of flow in a poppet valve. Proc 2st JHPS International Symposium on Fluid Power, Tokyo, 1993, 385–390

141   Watton J and Nelson RJ. Evaluation of an electrohydraulic forge valve behaviour using a CAD package. Applied Mathematical Modelling, Vol 17, 1993, 355–368

142   Davies RM and Watton J. Some practical considerations regarding the dynamic performance of proportional relief valves. 10th International Fluid Power Conference, Brugge, 1993, Mechanical Engineering Publications Ltd, 199–218.

143   Watton J and Bergada JM. Progress towards an understanding of the pressure/flow characteristics of a servovalve two flapper/double nozzle flow divider using CFD modelling. Proceedings of FLUCCME '94, Toulouse, France, 1994, 47–51 (ISBN 2-84088-010-5)

144 Lau KK and Edge KA. Impedance characteristics of hydraulic orifices. Proc IMechE Part I: Journal Systems and Control Engineering 1995, Vol 209, 241–253

145 Takashima M et al. Development of high performance components for pollution free water hydraulic system. Proc 3rd JHPS International Symposium on Fluid Power, Yokohama, 1996, 49–54

146 Helduser S and Muth A. Dynamic friction measurement method evaluated by means of cylinder and valves. Proc 3rd JHPS International Symposium on Fluid Power, Yokohama, 1996, 271–276

147 Handroos H and Halme J. Semi-empirical model of a counterbalance valve. Proc 3rd JHPS International Symposium on Fluid Power, Yokohama, 1996, 525–530

148 Hayashi S, Iizuka Y, Hayase T. Numerical analysis for stability of balanced piston type relief valve. Proc 3rd JHPS International Symposium on Fluid Power, Yokohama, 1996, 531–536

149 Palumbo A et al. Forces on a hydraulic spool valve. Proc 3rd JHPS International Symposium on Fluid Power, Yokohama, 1996, 543–548

150 Mokhtarzadeh-Dehghan MR, Ladommatos N, Brennan TJ. Finite element analysis of flow in a hydraulic pressure valve. Applied Mathematical Modelling 1997, Vol 21, 437–445

151 Washio S, Nakamura Y, Yu Y. Static characteristics of a piston-type pilot relief valve. Proc IMechE, Part C Journal of Mechanical Engineering Science, Vol 213, 1999, 231–239

152 Dong X and Ueno H. Flows and flow characteristics of spool valve. Proc 4th JHPS International Symposium on Fluid Power, Tokyo, 1999, 51–56

153 Maiti R, Surawattanawan P, Watton J. Performance prediction of a proportional solenoid control pressure relief valve. Proc 4th JHPS International Symposium on Fluid Power, Tokyo, 1999, 321–326

154 Nakanishi T et al. Numerical simulation of water hydraulic relief valve. Proc 4th JHPS International Symposium on Fluid Power, Tokyo, 1999, 555–560

155 Koivula T, Ellman A, Vilenius M. The effect of oil type on flow and cavitation properties in orifices and annular clearances. Power Transmission and Motion Control 1999, 151–165. Professional Engineering Publications Ltd

156 Borghi M, Milani M, Paoluzzi R. Stationary axial flow forces analysis on compensated spool valves. International Journal of Fluid Power, 2000, Vol 1, No 1, 17–25

157 Lee I-Y, Kim T-H, Kitagawa A. Technical problems on the linearisation of a servovalve in hydraulic control systems. Power Transmission and Motion Control 2001, 177–190. Professional Engineering Publications Ltd

158 Maiti R, Saha R, Watton J. The static and dynamic characteristics of a pressure relief valve with a proportional controlled pilot stage. Proc IMechE, Part I, Journal of Systems and Control Engineering, Vol 216,2002, 143–156. ISSN 0959–6518

159 Johnston DN, Edge KA, Brunelli M. Impedance and stability characteristics of a relief valve. Proc IMechE Vol 216 Part I: Journal Systems and Control Engineering 2002, 371–382

160 Chenvisuwat T, Park S, Kitagawa A. Development of a poppet type brake pressure control valve for a friction brake of rolling stock. 5th JFPS International Symposium on Fluid Power, Nara, 2002. Vol 3, 733–738

161 Yanping Hu, Deshum Liu. Static characteristics of relief valve with pilot G-π bridge hydraulic resistances network. Proc 5th JFPS International Symposium on Fluid Power, Nara, 2002, 739–744

162 Hayase T et al. Fundamental consideration on numerical analysis of unsteady flow through spool valve. Proc 5th JFPS International Symposium on Fluid Power, Nara, 2002, 929–934

163 Alirand M, Favennec G, Lebrun M. Pressure components stability analysis: a revisited approach. International Journal of Fluid Power, 2002, Vol 3, No 1, 33–46

164 Manhartsgruber B and Mikota J. Model-based parameter identification of a fluid power component. Power Transmission and Motion Control 2002, 229–244. Professional Engineering Publications Ltd

165 Zhang R, Alleyne AG, Presetiawan EA. Performance limitations of a class of two stage electrohydraulic flow valves. International Journal of Fluid Power, Vol 5, No 2, 2004, 47–53

166 Bergada JM and Watton J. A direct solution for flow rate and force along a cone-seated poppet valve for laminar flow conditions. Proc IMechE Part I: Journal Systems and Control Engineering 2004, Vol 218, 197–210

167 Gordic D, Babic M, Jovicic N. Modelling of spool position feedback servovalves. International Journal of Fluid Power, 2004, Vol 5, No 1, 37–50

168 Urata E. One-degree-of-freedom model for torque motor dynamics. International of Fluid Power, Vol 5, No 2, 2004, 35–42

169 Park S-H, Kitigawa A, Kawashima M. Water hydraulic high speed solenoid valve. Part 1: development and static behaviour. Proc IMechE Part I: Journal Systems and Control Engineering 2004, Vol 218, 399–409

Pumps, Motors and Associated Sub-Components

170 Wilson WE. Performance criteria for positive displacement pumps and fluid motors. Trans ASME 1949

171 Schlosser WMJ. Mathematical model for displacement pumps and motors. Hydraulic Power Transmission, 1961, April 252–257, & 269, May 324–328

172 Fisher, MJA theoretical determination of some characteristics of a tilted hydrostatic slipper bearing. B.H.R.A. Rep. RR 728 April 1962

173 Thoma JU. Mathematical models and effective performance of hydrostatic machines and transmissions. Hydraulic and Pneumatic Power, 179, Vol 15, 1969

174 Green WL and Crossley TR. An analysis of the control mechanism used in variable delivery hydraulic pumps. Proc IMechE, Fluid Plant and Machinery Group, Vol 185, 6/71, 1970–71, 63–72

175 Bown DE and Worton-Griffiths J. The dynamic characteristics of a hydrostatic transmission system. Proc IMechE, Fluid Plant and Machinery Group, Vol 186, 55/72, 1972, 755–773

176 Hibi A and Ichikawa T. Torque performance of hydraulic motor in whole operating condition from start to maximum speed and its mathematical model. Proc BHRA 4th International Fluid Power Symposium, Sheffield, England, 1975, B3 29–38

177 Martin MJ and Taylor R. Optimised port plate timing for an axial piston pump. 5th BHRA Fluid Power Symposium, Durham, England, 1978, B5–51–66

178 Hooke CJ and Kakoullis YP. The lubrication of slippers on axial piston pumps. 5th International Fluid Power Symposium September 1978, B2-(13–26) Durham, England

179 Hooke CJ and Kakoullis YP. The effects of centrifugal load and ball friction on the lubrication of slippers in axial piston pumps. 6th International Fluid Power Symposium, 179–191, Cambridge, England. 1981

180 McCandlish D and Dorey R. Steady state losses in hydrostatic pumps and motors. Proc 6th International Fluid Power Symposium, Cambridge, England, 1981, C3/133–144

181 Zarotti GL and Nervegna N. Pump efficiencies approximation and modelling. Proc 6th International Fluid Power Symposium, Cambridge, England, 1981, C4/145–164

182   Iboshi N and Yamaguchi A. Characteristics of a slipper bearing for swash plate type axial piston pumps and motors, theoretical analysis. Bulletin of the JSME, Vol 25, No 210, December 1982, 1921–1930

183   Iboshi N and Yamaguchi A. Characteristics of a slipper bearing for swash plate type axial piston pumps and motors, experiment. Bulletin of the JSME, Vol 26, No 219, September 1983. 1583–1589

184   Hooke CJ and Kakoullis YP. The effects of non flatness on the performance of slippers in axial piston pumps. Proc IMechE, December 1983, Vol. 197 C, 239–247

185   Watton J. Steady-state and dynamic flow characteristics of positive displacement pumps using laser doppler anemometry. Proceedings of the International Conference on Optical Techniques in Process Control. The Hague, June 1983, pp 165–178.

186   Seet G, Penny JE, Foster K. Application of a computer model in the design and development of a quiet vane pump. Proc IMechE Part B 199, 1985, 247–253

187   Foster K, Taylor R, Bidhendi IM. Computer prediction of cyclic excitation sources for an external gear pump. Proc IMechE Part C, Journal Mechanical Engineering Science, 199, B3, 1985, 175–180

188   Chapple PJ and Dorey RE. The performance comparison of hydrostatic piston motors-factors affecting their application and use. Proc 7th BHRA Fluid Power symposium, Bath, England, 1986, 1–8

189   Karmel AM. A study of the internal forces in a variable-displacement vane pump-Part 1: A theoretical analysis. Journal Fluids Engineering, 108, 1986, 227–232

190   Karmel AM. A study of the internal forces in a variable-displacement vane pump-Part 2: A parametric study. Journal Fluids Engineering, 108, 1986, 233–237

191   Hooke CJ and Li KY. The lubrication of overclamped slippers in axial piston pumps centrally loaded behaviour. Proc IMechE 1988, Vol 202, No C4, 287–293

192   Kobayashi, S, Hirose, M, Hatsue, J, Ikeya M. Friction characteristics of a ball joint in the swashplate type axial piston motor. Proc Eighth International Symposium on Fluid Power, J2–565–592, Birmingham, England, 1988

193   Hooke CJ and Li KY. The lubrication of slippers in axial piston pumps and motors. The effect of tilting couples. Proc IMechE 1989, Vol 203, part C, 343–350

194   Takahashi K and Ishizawa S. Viscous flow between parallel disks with time varying gap width and central fluid source. JHPS International Symposium on Fluid Power, Tokyo, March 1989, 407–414

195   Edge KA and Darling J. The pumping dynamics of swash plate piston pumps. Journal Dynamic Systems, Measurement and Control, 111, 1989, 307–312

196   Kobayashi S and Ikeya M. The structural analysis of piston balls and hydrostatic slipper bearings in swash plate type axial piston motors. Proc BHRA 9th International Symposium on Fluid Power, Cambridge, England, 1990, 19–32

197   Rampen WHS and Salter SH. The digital displacement pump. Proc BHRA 9th International Symposium on Fluid Power, Cambridge, England, 1990, 33–45

198   Watton J and Watkins-Franklin KL. The transient pressure characteristic of a positive displacement vane pump. Proc IMechE Journal Power and Energy, Vol 204 1990, 269–275

199   Li KY and Hooke C.J. A note on the lubrication of composite slippers in water based axial piston pumps and motors. Wear, 147 1991, 431–437

200   Johnston DN. Numerical modelling of reciprocating pumps with self-acting valves. Proc IMechE, Part I Journal of Systems and Control Engineering, Vol 205, 1991, 87–95

201   Koc E, Hooke CJ, Li KY. Slipper balance in axial piston pumps and motors.. Trans ASME, Journal of Tribology, Vol 114, Oct 1992, 766–772

202   Nishiumi T and Maeda T. The relationship between vane pump motion and chamber pressure in a vane pump. Proc 2st JHPS International Symposium on Fluid Power, Tokyo, 1993, 209–214

203   Zarotti LG and Paoluzzi R. Triple controls of variable displacement pumps. Proc 2st JHPS International Symposium on Fluid Power, Tokyo, 1993, 215–220

204   Manco S and Nervegna N. Pressure transients in an external gear hydraulic pump. Proc 2st JHPS International Symposium on Fluid Power, Tokyo, 1993, 221–227

205   Tanaka K, Nakahara T, Kyogoku K. Piston rotation and frictional forces between piston and cylinder of piston and motor. Proc 2nd JHPS International Symposium on Fluid Power, Tokyo, 1993, 235–240

206   Backe W and Kogl C. Secondary controlled motors in speed and torque control. Proc 2st JHPS International Symposium on Fluid Power, Tokyo, 1993, 241–248

207   Nagata K, Takahashi K, Saitoh K. A simulation technique for pressure fluctuations in a vane pump. 8th Bath International Fluid Power Workshop, Resaerch Studies Press Ltd, 1995, 169–183

208   Pettersson M, Weddfelt K, Palmberg J-O. Prediction of structural and audible noise from axial piston pumps using transfer functions. 8th Bath International Fluid Power Workshop, Research Studies Press Ltd. 1995, 184–203

209   Stecki JS and Chao L. Computer-aided techniques in contamination control. 8th Bath International Fluid Power Workshop, Research Studies Press Ltd, 1995, 349–363

210   Koc E, Hooke CJ. Investigation into the effects of orifice size, offset and oveclamp ratio on the lubrication of slipper bearings. Tribology International, Vol. 29, No 4, 299–305, 1996

211   Brookes CA et al. The selection and performance of ceramic components in a sea water pump. Proc 3rd JHPS International Symposium on Fluid Power, Yokohama, 1996, 3–12

212   Ito K, Kiyoshi I, Keiji S. Visualisation and detection of cavitation in V-shaped groove type valve plate of an axial piston pump. Proc 3rd JHPS International Symposium on Fluid Power, Yokohama, 1996, 67–72

213   Kosodo H et al. Experimental research about pressure-flow characteristics of V-notch. Proc 3rd JHPS International Symposium on Fluid Power, Yokohama, 1996, 73–78

214   Masuda K and Ohuchi H. Noise reduction of a variable piston pump with even number of cylinders. Proc 3rd JHPS International Symposium on Fluid Power, Yokohama, 1996, 91–96

215   You Z. Burton RT, Ukrainetz PR. Sliding mode control of a variable displacement pump. Proc 3rd JHPS International Symposium on Fluid Power, Yokohama, 1996, 169–176

216   Edge K, Burrows C, Lecky-Thomson N. Modelling of cavitation in a reciprocating plunger pump. Proc 3rd JHPS International Symposium on Fluid Power, Yokohama, 1996, 473–478

217   Koc E and Hooke C.J. Considerations in the design of partially hydrostatic slipper bearings. Tribology International, Vol 30, No. 11, 815–823, 1997

218   Ivantysynova M. A new approach to the design of sealing and bearing gaps of displacement machines. Proc 4th JHPS International Symposium on Fluid Power, Tokyo, 1999, 45–50

219   Gilardino L et al. An experience in simulation: the case of a variable displacement axial piston pump. Proc 4th JHPS International Symposium on Fluid Power, Tokyo, 1999, 85–91

220   Wieczoreck U and Ivantysynova M. CASPAR-A computer aided design tool for axial piston machines. Proc of the Power Transmission Motion and Control International Workshop, PTMC2000, Bath, UK. 2000, 113–126

221   Olems L. Investigations of the temperature behaviour of the piston cylinder assembly in axial piston pumps. International Journal of Fluid Power, 2000, Vol 1, No 1, 27–38

222   Manring ND and Zhang Y. The improved volumetric efficiency of an axial piston pump utilizing a trapped volume design. Trans ASME Journal Dynamic Systems, Measurement, and Control Vol 123, 2001, 479–487

223   Fairhurst M and Watton J. CFD analysis of a pump pressure compensator operating with a water based fluid. Power Transmission and Motion Control 2001, 21–32. Professional Engineering Publications Ltd

224   Ivantysynova M and Huang C. Investigation of the gap flow in displacement machines considering elastohydrodynamic effects. 5th JFPS International Symposium on Fluid Power, Nara, Japan, 2002, 219–229

225   Kazama T, Yamaguchi A, Fujiwara M. Motion of eccentrically and dynamically loaded hydrostatic thrust bearings in mixed lubrication. 5th JFPS 'International Symposium on Fluid Power, Nara, Japan, 2002, 233–238

226   Bergada JM and Watton J. A direct leakage flow rate calculation method for axial pump grooved pistons and slippers, and its evaluation for a 5/95 fluid application. 5th JFPS International Symposium on Fluid Power, Nara, Japan, 2002, 259–264

227   Iudicello F and Mitchell D. CFD modelling of the flow in a gerotor pump. Power Transmission and Motion Control 2002, 53–64. Professional Engineering Publications Ltd

228   Bergada JM and Watton J. Axial piston pump slipper balance with multiple lands. ASME International Mechanical Engineering Congress and exposition. IMECE 2002. New Orleans Louisiana November 17–22 2002. Vol 2 paper 39338

229   Ivantysynova M and Huang C. Investigation of the gap flow in displacement machines considering elastohydrodynamic effect. Proc 5th JFPS International Symposium on Fluid Power, Nara, 2002, 219–232

230   Manco S, Nervegna N, Gilardino L. Advances in the simulation of axial piston pumps. Proc 5th JFPS International Symposium on Fluid Power, Nara, 2002, 251–258

231   Bergada JM and Watton J. A direct leakage flow rate calculation method for axial piston grooved pistons and slippers, and its evaluation for a 5/95 fluid application. Proc 5th JFPS International Symposium on Fluid Power, Nara, 2002, 259–264

232   Yakabe S and Nagata K. Reduction of pressure fluctuation in a vane pump using genetic algorithm. Proc 5th JFPS International Symposium on Fluid Power, Nara, 2002, 271–276

233   Wieczorek U and Ivantysynova M. Computer aided optimization of bearing and sealing gaps in hydrostatic machines-the simulation tool CASPAR. International Journal of Fluid Power, 2002, Vol 3, No 1, 7–20

234   Manring ND and Kasaragadda SB. The theoretical flow ripple of an external gear pump. Trans ASME Journal Dynamic Systems, Measurement, and Control Vol 205, 2003, 396–404

235   Winkler B et al. Modelling and simulation of the elastohydrodynamic behaviour of sealing gaps. Proc 1st International Conference in Fluid Power Technology, Methods for Solving Practical Problems in Design and Control, Melbourne, Australia, 2003, 155–164. Fluid Power Net Publications

236   Kojima E. Development of a quieter variable displacement vane pump for automotive hydraulic power steering system. International Journal of Fluid Power, 2003, Vol 4, No 2, 5–14

237   Manring ND and Dong Z. The impact of using a secondary swash plate angle within an axial piston pump. Trans ASME Journal Dynamic Systems, Measurement, and Control Vol 126, 2004, 65–74

238   Ivantysynova M and Lasaar R. An investigation into micro and macrogeometric design of piston/cylinder assembly of swash plate machines. International Journal of Fluid Power, 2004, Vol 5, No 1, 23–36

239   Achten PAJ, van den Brink TL, Potma JW. Movement of the cups on the barrel plate of a floating cup axial piston machine. International Journal of Fluid Power, 2004, Vol 5, No 2, 25–33

240   Scharf S and Murrenhoff H. Measurement of friction forces between piston and bushing of an axial piston displacement unit. International Journal of Fluid Power, 6, 2000, No 1, 7–17

Fluid Compressibility

241   Smith LH, Peeler RL, Bernd LH. Hydraulic fluid bulk modulus-its effect on system performance and techniques for physical measurement. 16th National Conference on Industrial Hydraulics, USA, 1960, Vol 14. 179–196

242   Hayward ATJ. Aeration in hydraulic systems-its assessment and control. IMechE Proc Oil in Hydraulics Conference, 1961, 216–224

243   Green WL. The effects of discharge times on the selection of gas charged accumulators. Proc Third International Fluid Power Symposium, Turin, Italy 1973, D1/1–15

244   Bowns DE and Worton-Griffiths J. The effect of air in the fluid on the operating characteristics of a hydrostatic transmission. BHRA 3rd International Fluid Power Symposium, 1973

245   Paul FW, Walker AE, Robinson R. The effect of fluid line length on hydraulic gap controlled cold rolling mills. BHRA Proc Fluids in Control and Automation, Toronto, Canada, 1976, A6–115–131

246   Urata E. Influence of compressibility on the step response of a hydraulic servomechanism. Bulletin of the JSME, Vol 25, No 203, 1982, 797–803

247   Brault F. The use of hydraulic accumulators in energy storage. Proc 7th International Fluid Power Symposium, Bath 1986, 48/1–10

248   Yu J. Measurement of oil effective bulk modulus in hydraulic systems. No 3 Chinese Fluid Power Engineering 1991, 46–48

249   Johnston DN and Edge KA. In-situ measurement of the wave speed and bulk modulus in hydraulic systems. Proc IMechE 205 Part I Journal Systems and Control Engineering 1991, 191–197

250   Jen YM and Lee CB. Influence of an accumulator on the performance of a hydrostatic drive with control of the secondary unit. Proc IMechE 207 Part I Journal Systems and Control Engineering 1993, 173–184

251   Jinghong Y, Zhaoneng C, Yuanzhang L. The variation of oil effective bulk modulus in hydraulic systems. Trans ASME, Journal of Dynamics Systems, Measurement and Control, 1994, Vol 116, 146–150

252   Watton J and Xue. A new direct-measurement method for determining fluid bulk modulus in oil hydraulic systems. Proceedings FLUCOME '94 Toulouse, France, 1994, 543–548

253   Wang X et al. On the factors influencing the bubble content in air/hydro system. Proc 4th JHPS International Symposium on Fluid Power, Tokyo, 1999, 57–62

System Performance

254   Harper NF. Some considerations of hydraulic servos of jack type. Proc IMechE Conference on hydraulic servos, London, 1953, 41–50

255   Tou J and Sculthesis PM. Static and sliding friction in feedback systems. Journal of Applied Physics, 1953, 21, No 9, 1210–1217.

256   Shearer JL. Dynamic characteristics of valve-controlled hydraulic servomotors. Trans ASME, 1954, 895–903

257 Silberberg MY. A note on the describing function of an element with Coulomb, static and viscous friction. Trans AIEE, 1956, 75, Part 2, 423–425

258 Royle JK. Inherent non-linear effects of hydraulic control systems with inertia loading. Proc IMechE, 173, No9, 1959, 257–269

259 Turnbull DE. The response of a loaded hydraulic servomechanism. Proc IMechE, 173, No9, 1959, 270–284

260 Butler R. A theoretical analysis of the response of a loaded hydraulic relay. Proc IMechE, 173, No16, 1959, 429–455

261 Glaze SG. Analogue technique and the non-linear jack servomechanism. IMechE Automatic Control Conference, London, 1960, 178–188

262 Wang PKT and Ma JTS. Cavitation in valve controlled hydraulic actuators. Trans ASME Journal of Applied Mechanics, 1963, 537–546

263 Nikiforuk PN and Westland BE. The large signal response of a loaded high-pressure hydraulic servomechanism. Proc IMechE, 180, No32, 1965–6, 757–786

264 Carrington JE and Martin HR. Threshold problems in electro-hydraulic servomotors. Proc ImechE, Vol 180, PART 1, No 37, 1965–66, 881–894

265 Ashley T and Mills B. Frequency response of an electrohydraulic vibrator with inertia load. IMechE, Journal of Mechanical Engineering Science, 8, No1, 1966, 27–35

266 Martin HR and Lichtarowicz A. Theoretical investigation into the prevention of cavitation in hydraulic actuators. Proc IMechE, 181, No18, 1966–7, 423–431

267 Healey AJ and Stringer JD. Dynamic characteristics of an oil hydraulic constant speed drive. Proc IMechE, 183, 34, 1968–9, 682–692

268 Davies AM and Davies RM. Non-linear behaviour including jump resonance of hydraulic servomechanisms. IMechE, Journal of Mechanical Engineering Science, 11, 3, 1969, 837–846

269 Martin KF. Stability and step response of a hydraulic servo with special reference to unsymmetrical oil volume conditions. Proc IMechE Journal of Mechanical Engineering Science, 12, 1970, 331–338

270 Nikiforuk PN, Wilson JN, Lepp RM. Transient response of a time-optimised hydraulic servomechanism operating under cavitating conditions. Proc IMechE, 185, No60, 1970–71, 423–431

271 Guillon M and Blondel JP. Non-symmetrical cylinders and valves under non-symmetrical loading. Proc BHRA 2nd Fluid Power Symposium, Guildford, England, 1971, B5 85–111

272 Martin KF. Flow saturated response of a hydraulic servo. ASME Journal of Dynamic Systems, Measurement, and Control. 1974, 341–346

273 Watton J. The Dynamic steering behaviour and automation of a pipe laying machine. BHRA Proc Fluids in Control and Automation, Toronto, Canada, 1976, A1–1-32

274 Barnard BW and Dransfield P. Predicting response of a proposed hydraulic control system using bond graphs. Trans ASME Journal of Dynamic Systems, Measurement and Control, 1977, 1–8

275 LeVert FE. Dynamic analysis of a high speed electrohydraulic transient rod drive system. Fluidics Quarterly, 10, 2, 1978

276 Hilton DJ. Interactions between a pressure reducing valve and the upstream pipe. Proc 5th BHRA Fluid Power Symposium, Durham, England, 1978, G2–23–44

277 Takahashi K and Takahashi Y. Dynamic characteristics of a spool valve controlled servomotor with a non-symmetrical cylinder. Bull JSME,23, 1980, 1155–1162

278 Ikebe Y. Load insensitive electro-hydraulic servo system. The Journal of Fluid Control, 13, 4, 1981, 40–55

279 Vilenius MJ. The application of sensitivity analysis to electrohydraulic position servos. ASME Journal of Dynamic Systems, Measurement and Control 1983, 106, 77–82

280 Watton J. The generalised response of servovalve-controlled, single rod, linear actuators and the influence of transmission line dynamics. ASME Journal of Dynamic Systems, Measurement and Control,1984, 106, 157–162

281 Watton J. Closed-loop design of an electrohydraulic motor drive using open-loop steady-state characteristics. Journal of Fluid control, 20, 1, 1989, 7–30

282 Foster K and Fenney L. Characteristics and dynamic performance of electrical and hydraulic drives. Proc 1st JHPS International Symposium on Fluid Power, Tokyo, 1989, xvii-xxiv

283 Watton J. Optimum response design guides for electrohydraulic cylinder control systems. Journal of Applied Mathematical Modelling, Vol 14, 1990, 598–604

284 Watton J and Al-Baldawi RA. Performance optimisation of an electrohydraulic position control system with load-dependent supply pressure. Proc IMechE 205 Part I Journal of Systems and Control Engineering 1991, 175–189

285 Ramachandran S and Dransfield P. Modeling, analysis and simulation of an electro-hydraulic flight control actuation system including friction. Proc 2st JHPS International Symposium on Fluid Power, Tokyo, 1993, 203–208

286 Zhao T and Virvalo T. Fuzzy state controller and its application in hydraulic position servo. Proc 2nd JHPS International Symposium on Fluid Power, Tokyo, 1993, 417–422

287 Burton JD, Edge KA, Burrows CR. Analysis of an electro-hydraulic position control system using transmission line modelling (TLM). Proc 2st JHPS International Symposium on Fluid Power, Tokyo, 1993, 507–514

288 Davies RM and Watton J. Intelligent control of an electrohydraulic motor drive system. Journal of Mechatronics, Vol 5, No 5, 1995, 527–540

289 Watton J and Xue Y. Identification of fluid power component behaviour using dynamic flow rate measurement. Proc IMechE Vol 209 Journal Systems and Control Engineering, 1995, 179–191

290 Xue Y and Watton J. A self-organising neural network approach to data-based modelling of fluid power systems dynamics using the GMDH algorithm. Proc IMechE Journal Systems and Control Engineering, 1995, Vol 209, 229–240

291 Sidders JA, Tilley DG, Chapple PJ. Thermal-hydraulic performance prediction in fluid power systems. Proc IMechE Part I Journal Systems and Control Engineering, Vol 210, 1996, 231–242

292 Xu L, Schueller JK, Harrell R. Dynamic response of a servovalve controlled hydraulic motor driven centrifugal pump. Trans ASME Journal of Dynamic Systems, Measurement and Control, Vol 118, 1996, 253–258

293 Koskinen KT, Makinen E, Vilenius MJ, Virvalo T. Position control of a water hydraulic cylinder. Proc 3rd JHPS International Symposium on Fluid Power, Yokohama, 1996, 43–48

294 Yamashina C, Miyakawa S, Urata E. Development of water hydraulic cylinder position control system. Proc 3rd JHPS International Symposium on Fluid Power, Yokohama, 1996, 55–60

295 Matsui T and Koseki H. Motion control of water hydraulic servo system. Proc 3rd JHPS International Symposium on Fluid Power, Yokohama, 1996, 61–66

296 Suzuki K and Tomioka K. Improving characteristics of electrohydraulic servo system with nonsymmetrical cylinder using DSP. Proc 3rd JHPS International Symposium on Fluid Power, Yokohama, 1996, 201–206

297 Watton J and Kwon K-S. Neural network modelling of fluid power control systems using internal state variables. Mechatronics, 1996, Vol 6, No 7, 817–827

298 Nishiumi T and Watton J. Model reference adaptive control of an electrohydraulic motor drive using an artificial neural network compensator. Proc IMechE, Part I, Journal of Systems and Control Engineering, 1997, Vol 216, 357–367

299   Watton J and Xue Y. Simulation of fluid power circuits using artificial network models, Part 1: selection of component models. Proc IMechE, Part I: Journal of Systems and Control Engineering, 1997, Vol 211, 111–122

300   Xue Y and Watton J. Simulation of fluid power circuits using artificial network models, Part 2: circuit simulation. Proc IMechE, Part I: Journal of Systems and Control Engineering, 1997, Vol 211, 429–438

301   Xue Y and Watton J. Dynamics modelling of fluid power systems applying a global error descent algorithm to a self-organising radial basis function network. Mechatronics, 1998, No 8, 727–745

302   Scheidl R and Manhartsgruber B. On the dynamic behaviour of servo-hydraulic drives. Nonlinear Dynamics, 17, 1998, 247–268

303   Suzuki K, Sugi S, Ueda H. Improving the characteristics of and electrohydraulic servo system with nonsymmetrical cylinder by ZPETC and linearisation. Proc 4th JHPS International Symposium on Fluid Power, Tokyo, 1999, 93–98

304   Pai K-R and Shih M-C. Multi-speed control of a hydraulic cylinder using self-tuning fuzzy control method. Proc 4th JHPS International Symposium on Fluid Power, Tokyo, 1999, 99–104

305   Virvalo T. On the motion control of a hydraulic servo cylinder drive. Proc 4th JHPS International Symposium on Fluid Power, Tokyo, 1999, 105–110

306   Sato Y and Tada K. Rotational speed control of a hydraulic servomotor with a large inertia load using sliding mode control. Proc 4th JHPS International Symposium on Fluid Power, Tokyo, 1999, 119–124

307   Baum H and Murrenhoff H. Use of neural networks for the simulation of hydraulic systems including fluid temperature dependent component efficiencies. Power Transmission and Motion Control 2001, 57–71. Professional Engineering Publications Ltd

308   Virvalo T. Comparing controllers in hydraulic motion control. Power Transmission and Motion Control 2002, 215–228. Professional Engineering Publications Ltd

309   Vilenius M, Koskinen KT, Lakkonen M. Water hydraulics motion control-posibilities and challenges. Proc 5th JFPS International Symposium on Fluid Power, Nara, 2002, 3–10

310   Lisowski E and Czyzycki W. Simulation of vane pump controller by use of the DELPHI software environment. Proc 1st International Conference in Fluid Power Technology, Melbourne, Australia, 2003, 261–270. Published by Fluid Power Net Publications

311   Noskievic P. Simulation and dynamic analysis of mechatronic systems with the output in the virtual reality world. Proc 1st International Conference in Fluid Power Technology, Melbourne, Australia, 2003, 543–551. Published by Fluid Power Net Publications

312   Sampson E, Habibi S, Burton R, Chinniah. Effect of controller in reducing steady state error due to flow and force disturbances in the electrohydraulic actuator system. International J of Fluid Power, 2004, Vol 5, No 2, 57–66

Transmission Lines

313   Iberall AS. Attenuation of oscillatory pressures in instrument lines. Journal of Research of the National Bureau of Standards, 1950, Vol 45, rp 2115

314   Rohmann CP and Grogan EC. On the dynamics of pneumatic transmission lines. Trans ASME, 1957, 853–874

315   Nichols NB. The linear properties of pneumatic transmission lines. Trans Instrument Society of America, 1962, 5–14

316   Brown FT. Transient response of fluid lines. Trans ASME Journal of Basic Engineering, 1962, Vol 84, 547–553

317   Ansari JS and Oldenburger R. Propagation of disturbance in fluid lines. Trans ASME, Journal of Basic Engineering, 1967, 415–452

318   Karam JT and Franke ME. The frequency response of pneumatic lines. Trans ASME, Journal of Basic Engineering, 1967, 371–377

319   Zielke W. Frequency dependent friction in transient pipe flow. Trans ASME, Journal of Basic Engineering, 1968, 109–115

320   Karam J T and Leonard R G. A simple yet theoretically based model for simulating fluid transmission line systems. ASME Journal of Basic Engineering, 1973, 498–504

321   Trikha A K. An efficient method for simulating frequency dependent friction in transient liquid flow. ASME Journal of Fluids Engineering, 1975, 97–104

322   Katz S. Transient response of fluid lines by frequency response conversion. Trans ASME, Journal of Dynamic Systems, Measurement and Control, 1977, 311–313

323   Muto T and Kanei T. Resonance and transient response of pressurised complex systems. Bulletin JSME, 1980, Vol 23, No 184, 1610–1617

324   Hullender D A and Healey A J. Rational Polynomial approximation for fluid transmission line models. Fluid Transmission Line Dynamics, published by ASME, 1981, 33–56

325   Tanahashi T. Distorted pressure histories due to the step response in a linear tapered line. Bulletin JSME, 1982, Vol 25, No 208, 1521–1528

326   Hsue C Y and Hullender D A. Modal approximations for the fluid dynamics of hydraulic and pneumatic transmission lines. Fluid Transmission Line Dynamics, Published by ASME, 1983, 51–77

327   Stecki J S and Davis D C. Fluid transmission lines-distributed parameter models. Part 1: a review of the state of the art. Proc IMechE, 200, 1986, Part A, 215–228

328   Stecki J S and Davis D C. Fluid transmission lines-distributed parameter models. Part 2: comparison of models. Proc IMechE, 200, 1986, Part A, 229–236

329   Kitsios E E and Boucher R F. Transmission line modelling of a hydraulic position control system. Proc IMechE, 1986, 200, No B4, 229–236

330   Johnston DN and Drew JE. Measurement of positive displacement pump flow ripple and impedance. Proc IMechE, Part I, Journal of Systems and Control Engineering, vol 210, 1996, 65–74

331   Watton J. The dynamic performance of an electrohydraulic servovalve/motor system with transmission line effects. ASME J Dynamic Systems, Measurement and Control, 1987, 109, 14–18

332   Watton J and Tadmori M J. A comparison of techniques for the analysis of transmission line dynamics in electrohydraulic control systems. Journal of Applied Mathematical Modelling, 1988, 12, 457–466

333   Watton J. Modelling of electrohydraulic systems with transmission lines using modal approximations. Proc IMechE, 1988, 202, No B3, 153–163.

334   Suzuki K, Taketomi T, Sato S. Improving Zielke's method of simulating frequency-dependent friction in laminar liquid pipe flow. Trans ASME, Journal of Fluids Engineering, 1991, Vol 113, 569–573

335   Yang WC and Tobler WE. Dissipative modal approximation of fluid transmission lines using linear friction models. Trans ASME, J of Dynamic Systems, Measurement and Control, 1991, Vol 113, No 1, 152–162

336   Sanada K, Richards C W, Longmore D K, Johnstone D N. A finite element model of hydraulic pipelines using an optimized interlacing grid system. Proc IMechE, Journal of Systems and Control Eng, 1993, 207, 213–222

337   Krus P, Weddfelt K, Palmberg J-O. Fast pipeline models for simulation of hydraulic systems. Trans ASME, 1994, 116,132–136

338   Burton J D, Edge K A, Burrows C R. Modelling requirements for the parallel simulation of hydraulic systems. Proc IMechE, J Systems, Measurement, and Control, 1994, 116, 137–145

339   Piche R and Ellman A. A standard hydraulic fluid transmission line model for use with ODE simulators. Proc 8th Bath International Fluid Power Workshop, published by Research Studies Press, 1996, 221–236

340   Watton J and Hawkley C J. An approach for the synthesis of oil hydraulic transmission line dynamics utilising in situ measurements. Proc IMechE, Journal of Systems and Control Engineering, 1996, 210, 77–93

341   Sanada K and Kitugawa A. A study on H$_\infty$ control of a closed loop pressure control system considering pipeline dynamics. Proc 3rd JHPS International Symposium on Fluid Power, Yokohama, 1996, 177–182

342   Taylor SEM, Johnston DN, Longmore DK. Modelling of transient flow in hydraulic pipelines. Proc IMechE, Journal of Systems and Control Engineering, 1997, 211, 447–456

343   Kannisto S and Virvalo T. Hydraulic pressure in long hose. Power Transmission and Motion Control 2002, 165–176. Professional Engineering Publications Ltd

344   Kojima E and Shinada M. Development of accurate and practical simulation technique based on the modal approximations for fluid transients in compound fluid line systems. 1st report: establishment of fundamental calculation algorithm and basic considerations for verification of its availability. International Journal of Fluid Power, 2002, Vol 4, No 2, 5–15

345   Shinada M and Kojima E. Development of a practical and high accuracy simulation technique based on numerical modal approximation for fluid transients in compound fluid line systems. Proc 5th JFPS International Symposium on Fluid Power, Nara, 2002, 871–876

346   Kojima E and Shinada M. Development of accurate and practical simulation technique based on the modal approximations for fluid transients in compound fluid line systems. 2nd report: enhancement of analytic functions for generalization. International Journal of Fluid Power, 2003, Vol 4, No 3, 35–45

347   Manhartsgruber B. Passivity of fluid transmission line models. Power Transmission and Motion Control Workshop, PTMC 2004, Bath, 99–108. Published by Professional Engineering Publications Ltd

Fluids, Oil analysis, Wear and Noise

348   Lichtarowicz A. Cavitating jet apparatus for cavitation erosion test. ASTM STP 664, 1979, 530–549

349   Tsai CP. Particle counting in water based fluids using light blockage type automatic instruments. 6th International Fluid Power Conference, Cambridge, UK, 1981, 87–94

350   Hunt TM and Tilley DG. Techniques for the assessment of contamination in hydraulic oils. IMechE Contamination Control in Hydraulic Systems, Bath, UK, 1984, 57–63

351   McCullagh P. Ferrography and particle analysis in hydraulic power systems. IMechE Contamination Control in Hydraulic Systems, Bath, UK, 1984, 65–77

352   Braun S (ed.). Mechanical Signature Analysis.. Academic Press, 1986

353   Day MJ, Way NR, Thompson K. The use of particle counting techniques in the condition monitoring of fluid power systems. Condition Monitoring '87, edited by MH Jones. Pineridge Press Ltd, 1987, 322–339

354   Raw I. Particle size analyser based on filter blockage. Condition Monitoring '87, edited by MH Jones, Pineridge Press Ltd, 1987, 875–894

355   Lewis RT. Analysis of ferrous wear debris. Condition Monitoring '87, edited by MH Jones, Pineridge Press Ltd, 1987, 360–370

356 Price AL, Roylance BJ, Zie LX. The PQ-a method for the rapid quantification of wear debris. Condition Monitoring '87, edited by MH Jones, Pineridge Press Ltd, 1987, 391–405

357 Silva G. Wear generation in hydraulic pumps. SAE International Off-Highway and Powerplant Conference, Milwaukee USA, 1990, paper 901679

358 Bergmemann M. Noise problems of hydraulic piston pumps with odd and even numbers of cylinders. Proc BHRA 9th International Symposium on Fluid Power, Cambridge, England, 1990, 235–248

359 Rinkinen. J., Kiiso, T. Using portable particle counter in oil system contamination control. The 3rd Scandinavian International Conference on Fluid Power, Linköping, Sweden, 1993, Vol. 1, 309–328

360 Ramden T, Weddfelt K, Palmberg Jan-Ove. Condition monitoring of fluid power pumps by vibration measurement. 10th International Conference on Fluid Power-The future for Hydraulics, Brugge, Belgium, MEP Publications Ltd, 1993, 263–276

361 Sommer HT, Raze TL, Hart JM. The effects of optical material properties on particle counting results of light scattering and extinction sensors. 10th International Conference on Fluid Power-The future for Hydraulics, Brugge, Belgium, MEP Publications Ltd, 1993, 289–308

362 Ramden T. On Condition Monitoring of Fluid Power Pumps and Systems. Linkoping Studies in Science and Technology, Thesis No 474, Linkoping University, 1995

363 Hodges P. Hydraulic Fluids. Arnold, 1996

364 Ichiyanagi T, Kojima E, Edge KA. The fluid borne noise characteristics of a bent axis motor established using the 'Secondary Source' method. Proc 5th Scandinavian Conference on Fluid Power, Linköping, Sweden, 1997, 123–138

365 Sandt J, Rinkinen J, Laukka J. Particle and water on-line monitoring for hydraulic system diagnosis. Proc 5th Scandinavian Conference on Fluid Power, Linköping, Sweden, 1997, 257–268

366 Stecki JS (ed).Total Contamination Control. Fluid Power Net, 1998

367 Urata E. Cavitation erosion in various fluids. Power Transmission and Motion Control 1998, 269–284. Professional Engineering Publications Ltd

368 Kwong AHM and Edge KA. A method to reduce noise in hydraulic systems by optimizing clamp locations. IMechE Part I Journal of Systems and Control Engineering, Vol 212 No 14, 1998, 267–280

369 Martin KF and Marzi MH. Diagnostics of a coolant system via neural networks. Proc IMechE Part I Journal of Systems and Control Engineering, Vol 213, 1999, 229–241

370 Roylance BJ and Hunt TM. Wear Debris Analysis. Coxmoor, 1999

371 Environmental impact of fluid power systems. Organised by the IMechE, London, 1999, collection of 11 papers

372 Edge KA. Designing quieter hydraulic systems-some recent developments and contributions. Proc 4th JHPS International Symposium on Fluid Power, Tokyo, 1999, 3–27

373 Totten GE, Kling GH, Reichel J. Development of hydraulic pump performance standards an overview of current activities. Proc 4th JHPS International Symposium on Fluid Power, Tokyo, 1999, 63–67

374 Ohuchi H and Masuda K. Active noise control of a variable displacement axial piston pump with even number of cylinders. Proc 4th JHPS International Symposium on Fluid Power, Tokyo, 1999, 79–84

375 Totten GE, Reichel J, Kling GH. Biodegradable fluids: a review. Proc 4th JHPS International Symposium on Fluid Power, Tokyo, 1999, 285–290

376 Li ZY et al. The development and perspective of water hydraulics. Proc 4th JHPS International Symposium on Fluid Power, Tokyo, 1999, 335–342

377   Iudicello F and Baseley S. Fluid-borne noise characteristics of hydraulic and electro-hydraulic pumps. Power Transmission and Motion Control 1999, 313–323. Professional Engineering Publications Ltd

378   Qatu M et al. Analytical versus test-bench results for fluid borne noise simulation. Power Transmission and Motion Control 1999, 324–334. Professional Engineering Publications Ltd

379   Stecki JS (ed.). Total Contamination Control. Fluid Power Net, 2000

380   Holroyd T. Acoustic emission & ultrasonics.. Published by Coxmoor, 2000

381   Tikkanen S. Influence of line design on pump performance. Power Transmission and Motion Control 2001, 33–46. Professional Engineering Publications Ltd

382   Shao Y and Nezu K. Extracting symptoms of bearing faults from noise using a non-linear filter. Proc IMechE, Part I, Journal of Systems and Control Engineering, 2002, Vol 216, 169–179

383   Peters RJ. Noise and acoustics.. Published by Coxmoor, 2002

384   Riipinen H et al. Effects of microbial growth and particles on filtration in water hydraulic systems. Proc 5th JFPS International Symposium on Fluid Power, Nara, Japan, 2002, 173–176

385   Suzuki K and Urata E. Cavitation erosion of materials for water hydraulics. Power Transmission and Motion Control 2002, 127–139. Professional Engineering Publications Ltd

386   Ijas M and Virvalo T. Problems in using an accumulator as a pressure damper. Power Transmission and Motion Control 2002, 277–289. Professional Engineering Publications Ltd

387   Urata E. Evaluation of filtration performance of a filter element. Power Transmission and Motion Control 2002, 291–304. Professional Engineering Publications Ltd

388   Johansson A, Andersson J, Palmberg J-O. Optimal design of the cross-angle for pulsation reduction in variable displacement pumps. Power Transmission and Motion Control 2002, 319–333. Professional Engineering Publications Ltd

389   Johansson A and Palmberg J-O. Quieter hydraulic systems-design considerations. Proc 5th JFPS International Symposium on Fluid Power, Nara, Japan, 2002, 799–804

390   Gao H et al. Numerical and experimental investigation of cavitating flow in oil hydraulic ball valve. Proc 5th JFPS International Symposium on Fluid Power, Nara, Japan, 2002, 923–928

391   Radhakrishnan M. Hydraulic Fluids. ASME Press N York, 2003

392   Nikkila P anf Vilenius M. The simulation of cleanliness level in hydraulics. Proc 1st International Conference in Fluid Power Technology, Methods for Solving Practical Problems in Design and Control, Melbourne, Australia, 2003, 233–244. Fluid Power Net Publications

393   Rinkinen, J., Ahlstedt, H. & Nurminen, T. 2003. Experiences of fluid flow modelling in the design of mobile hydraulic oil tanks. Stecki, J.S. (Ed.). 1st International Conference on Computational Method in Fluid Power Technology, Methods for Solving Practical Problems in Design and Control, Melbourne, Australia, 2003,123–136

394   Evans JS and Hunt TM. Oil analysis. Coxmoor, 2004

395   Scheffer C and Girdhar P. Practical Machinery Vibration Analysis & Predictive Maintenance. Elsevier 2004

396   Gohler O-C, Murrenhoff H, Schmidt M. Ageing simulation of biodegradable fluids by means of neural networks. Power Transmission and Motion Control Workshop, PTMC 2004, Bath, 71–83. Published by Professional Engineering Publications Ltd

397   Vael GEM, Lopez I, Achten PAJ. Reducing noise pulsation with the floating cup pump-theoretical analysis. Power Transmission and Motion Control Workshop, PTMC 2004, Bath, 123–141. Published by Professional Engineering Publications Ltd

Author Biography

John Watton is Professor of Fluid Power in the School of Engineering at Cardiff University and has worked continually in fluid power for the past 36 years. He has worked in industry as a draughtsman and a Systems Engineer, has maintained a strong link with the fluid power industry via consultancy and commercial R&D, and has significant experience as an Expert Witness. He has been involved as co-designer of three novel mobile machines for general loader/Agricultural/Forestry and Civil Engineering applications. The work on condition monitoring and fault diagnosis, and in particular the application of on-line computer-based techniques, has led to pioneering applications in the steel processing industry. He has published over 160 papers on a wide range of fluid power topics such as component and systems modelling, computer control, fluid mechanics of servovalves/pumps/motors, dynamic performance and control of forging presses, automotive steering and suspensions, and has supervised many PhD students. Professor Watton teaches Control Theory, Condition Monitoring, Fluid Power, Fluid Mechanics, Vibrations, at the undergraduate level.

Index

Printing: Krips bv, Meppel
Binding: Stürtz, Würzburg